second edition

Copymasters

Blue Level

ROSE GRIFFITHS

Heinemann Educational Publishers
Halley Court, Jordan Hill, Oxford, OX2 8EJ
a division of Harcourt Education Ltd
www.myprimary.co.uk

Heinemann is a registered trademark of Harcourt Education Ltd

First edition first published 1996

Second edition first published 2005

10 09 08 87 06 05
10 9 8 7 6 5 4 3 2 1

ISBN 0 435 02144 3

Designed and typeset by Susan Clarke
Illustrated by Mick Reid
Cover design by Susan Clarke
Repro by Digital Imaging, Glasgow
Printed and bound in the UK by Thomson Litho

The author and publishers would like to thank teachers at the following schools for their help in trialling these materials:
Murrayburn Primary School, Edinburgh
St Bernard's RC Primary School, Bristol
Plymouth Grove Primary School, Manchester
Lubenham Primary School, Leicester
St Peters CE Primary School, Leicester
St Peters CV Primary School, Gwent
Cawley Lane Junior School, West Yorkshire
Naphill and Walters Ash Combined School, High Wycombe
Emmer Green Primary School, Reading
St Anne's JMI School, Streetly

Contents

Part 1		
Fill the box	Counting to 50	B1
More or less	Counting backwards or forwards to 50	B2, B3
Spelling numbers	Spelling one to ten	B4, B5
Eight bats	Addition bonds to 8	B6, B7
Number links	Links between addition and subtraction	B8, B9
Off by heart	Mental recall of addition within 8	B10
Spiders and snakes	Multiples of 2 and 10 to 50	B11, B12
Speedy sums A/Speedy sums B	Mental recall of bonds within 8	B13
Speedy sums C/Speedy sums D	Mental recall of bonds within 8	B14
T-shirts	Addition and subtraction within 20	B15, B16
Bowling	Addition to 10	B17
Fives and ones	Counting in 5s to 50	B18, B19
Pick up bricks	Grouping in 10s to 50	B20, B21
Bat and fives	Multiples of 5 to 50	B22, B23
Card sums	Addition within 20	B24, B25
Sums which make 8 game	Mental recall of addition within 8	B26, B27, B28
Fifty pence game	Counting in 5s to 50	B29, B30 **GP**

Part 2		
Coins in a jar	Counting to 60	B31
Sums in words	Spelling eleven to twenty	B32, B33
Tens and ones	Counting in 10s and 1s to 60	B34, B35
Two times table	Two times table	B36, B37
Hours and half hours	Using halves; telling the time	B38, B39
Nine counters	Addition bonds to 9	B40, B41
Fives and tens	Using 5s and 10s to 50	B42, B43
Speedy sums E/Speedy sums F	Mental recall of bonds within 9	B44
Boxes	Addition to 20	B45, B46
Teen numbers	Addition within 20	B47, B48
Five times table	Fives times table	B49, B50
More teen numbers	Subtraction within 20	B51
Photos	Addition and subtraction within 24	B52, B53
Hopping frogs	Multiples of 3 to 18	B54, B55

Sixty pence game	Counting to 60	B56, B57 **GP**
Tens and ones game	Counting in 10s and 1s to 60	B58, B59 **GP**
Make 9 game	Addition bonds to 9	B60, B61

Part 3

Joke shop	Counting to 75	B62
Tens and teens	Spelling thirty to seventy	B63, B64
What comes next?	Multiples of 2, 5 and 10	B65, B66
Footballs	Addition and subtraction bonds to 10	B67, B68
Easier adding	Addition within 30	B69, B70
Speedy sums G/Speedy sums H	Mental recall of bonds within 10	B71
Adding up	Addition within 40	B72, B73
Three times table	Three times table	B74, B75
Fifty pences	Using 50p coins and amounts over £1	B76, B77
Dog's toys	Multiplication and division by 2	B78, B79
Swimming	Mixed problems	B80, B81
Dog food	Multiplication and division by 2	B82, B83
Taking away	Subtraction within 40	B84, B85
Secret numbers	Number properties	B86, B87
Tens and teens bingo	Saying and listening to tens and teen numbers	B88, B89 **GP**
Sums which make 10 game	Addition and subtraction bonds to 10	B90, B91, B92 **GP**

GP Colour versions of these games are included in the *Games Pack*.

Fill the box

How many paperclips? _____

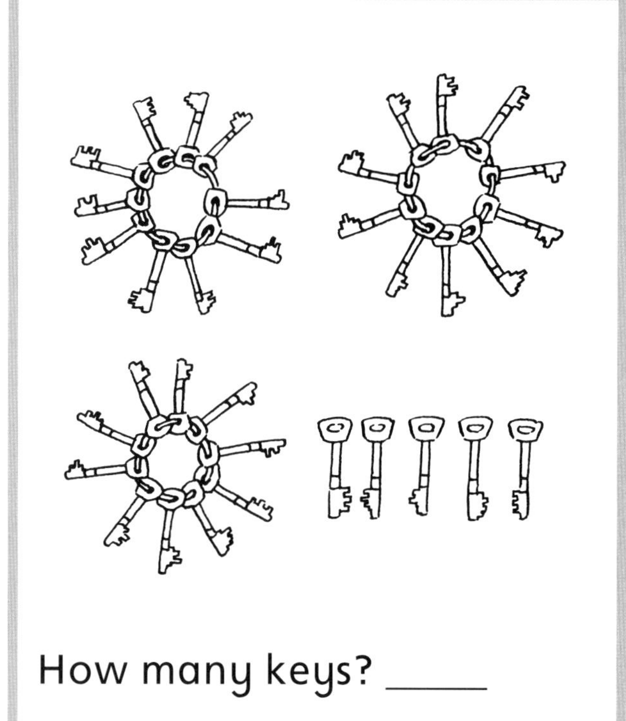

How many keys? _____

Draw the missing things.

40 candles

49 seeds

More or less

Name _______________________

Date _______________________

Join the dots.
Start from 40, and work backwards.
40, then 39, then 38, …

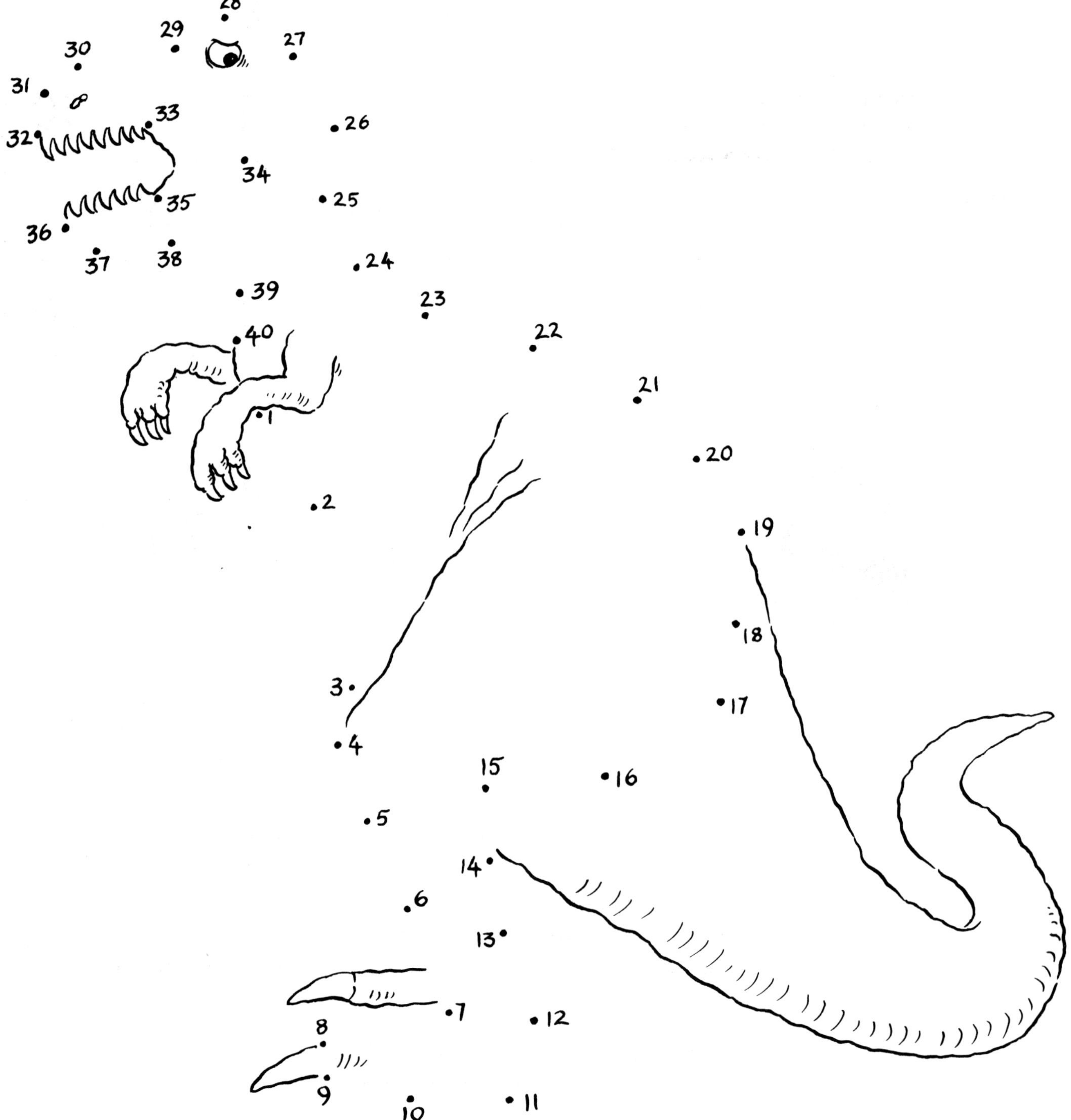

More or less

Name _____________________

Date _____________________

Fill in the missing numbers.

1	2	3	4			7		9	10
11	12			15	16		18		
	22	23	24				28		30
31		33		35		37		39	
	42		44		46		48		50

What is one less than 20? _______

What is one less than 30? _______

What is one less than 40? _______

What is one less than 50? _______

Spelling numbers

Name ______________________

Date ______________________

1

one
on__
o___
____
one

2

two
tw__
t___
____
two

3

three
th___
th___
__ree
__ree
____
three

4

four
f___
f___
f__r
_our
____
four

5

five
f___
f___
fi__
_ive
____
five

Mixed up numbers! Sort them out.

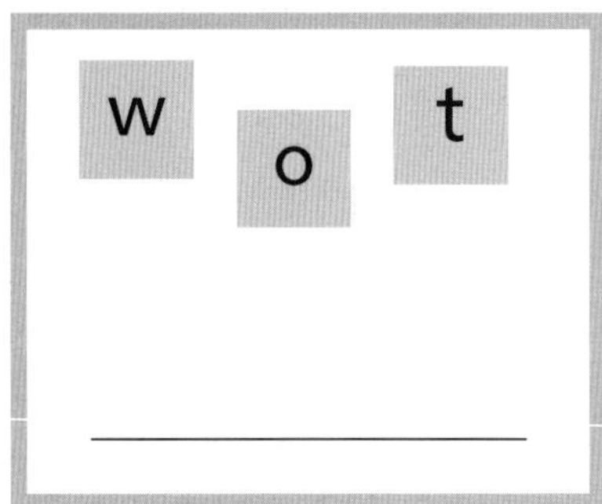

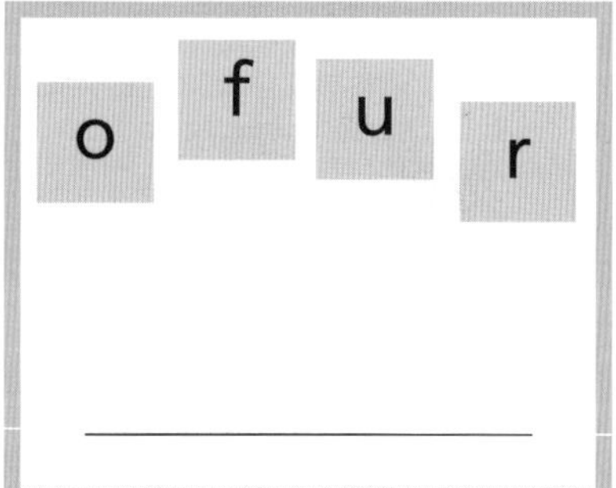

____   ____   ____

Spelling numbers

Name _______________________

Date _______________________

6

six
si__
s___
___ ___
six

10

ten
te__
t___
___ ___
ten

7

seven
se____
se____
___ven
___ven
____
seven

8

eight
ei____
ei____
__ght
__ght
____
eight

9

nine
ni___
ni___
n____
n____
____
nine

Mixed up numbers! Sort them out.

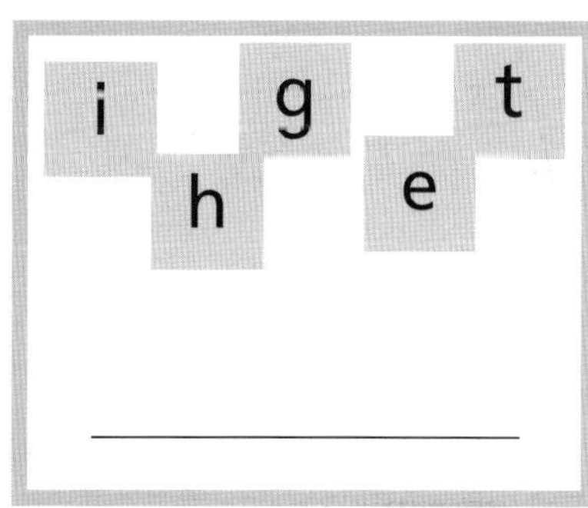

____

____

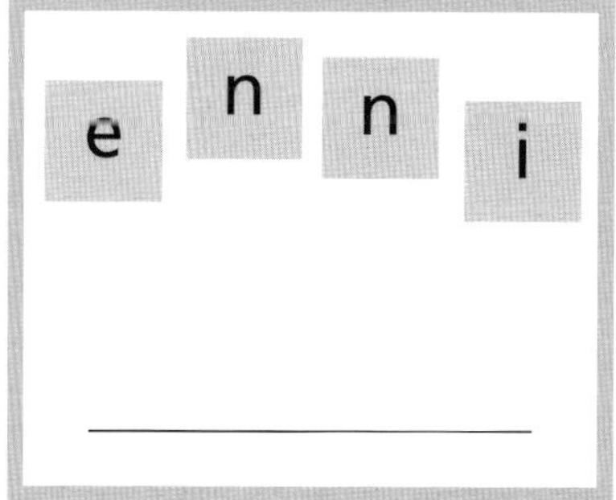

____

Eight bats

Name _______________________

Date _______________________

Cut out the numbers at the bottom.

2	+		=	8
3	+		=	8
0	+		=	8
6	+		=	8
1	+		=	8
5	+		=	8
7	+		=	8
4	+		=	8

✂ -

| 1 | 2 | 3 | 4 | 5 | 6 | 7 | 8 |

Eight bats

Name _______________________

Date _______________________

Fill in the missing numbers.

3 + ☐ = 8	☐ + 3 = 8
1 + ☐ = 8	☐ + 7 = 8
5 + ☐ = 8	☐ + 0 = 8
2 + ☐ = 8	☐ + 4 = 8
7 + ☐ = 8	☐ + 6 = 8
4 + ☐ = 8	☐ + 2 = 8
6 + ☐ = 8	☐ + 5 = 8

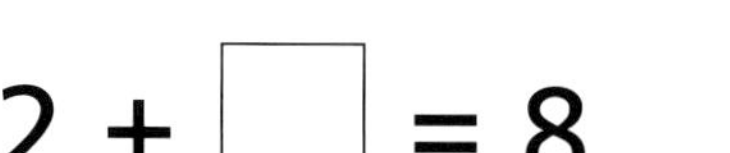

2 + ☐ = 8	6 + ☐ = 8
☐ + 2 = 8	☐ + 6 = 8

3 + ☐ = 8	5 + ☐ = 8
☐ + 3 = 8	☐ + 5 = 8

Number links

Name ______________________

Date ______________________

Use bricks to do these.

1   6   7

$1 + 6 = \underline{}$

$6 + 1 = \underline{}$

$7 - 1 = \underline{}$

$7 - 6 = \underline{}$

2   6   8

$2 + 6 = \underline{}$

$6 + 2 = \underline{}$

$8 - 2 = \underline{}$

$8 - 6 = \underline{}$

3   5   8

$3 + 5 = \underline{}$

$5 + 3 = \underline{}$

$8 - 3 = \underline{}$

$8 - 5 = \underline{}$

1   4   5

$1 + 4 = \underline{}$

$4 + 1 = \underline{}$

$5 - 1 = \underline{}$

$5 - 4 = \underline{}$

2   4   6

$\underline{} + \underline{} = \underline{}$

$\underline{} + \underline{} = \underline{}$

$\underline{} - \underline{} = \underline{}$

$\underline{} - \underline{} = \underline{}$

Number links

Name _______________________

Date _______________________

5 + 3 = ___
3 + 5 = ___
8 − 5 = ___
8 − 3 = ___

1 + 5 = ___
5 + 1 = ___
6 − 5 = ___
6 − 1 = ___

3 + 4 = ___
4 + 3 = ___
7 − 3 = ___
7 − 4 = ___

5 + 2 = ___
2 + 5 = ___
7 − 2 = ___
7 − 5 = ___

2 + 6 = ___
6 + 2 = ___
8 − 2 = ___
8 − 6 = ___

2 + 4 = ___
4 + 2 = ___
6 − 2 = ___
6 − 4 = ___

3 + 3 = ___
6 − 3 = ___

1 + 7 = ___
7 + 1 = ___
8 − 1 = ___
8 − 7 = ___

4 + 4 = ___
8 − 4 = ___

Off by heart

Name _______________________

Date _______________________

1 + 0 = ___

1 + 1 = ___

1 + 2 = ___

1 + 3 = ___

1 + 4 = ___

1 + 5 = ___

1 + 6 = ___

1 + 7 = ___

2 + 0 = ___

2 + 1 = ___

2 + 2 = ___

2 + 3 = ___

2 + 4 = ___

2 + 5 = ___

2 + 6 = ___

3 + 0 = ___

3 + 1 = ___

3 + 2 = ___

3 + 3 = ___

3 + 4 = ___

3 + 5 = ___

6 + 0 = ___

6 + 1 = ___

6 + 2 = ___

5 + 0 = ___

5 + 1 = ___

5 + 2 = ___

5 + 3 = ___

4 + 0 = ___

4 + 1 = ___

4 + 2 = ___

4 + 3 = ___

4 + 4 = ___

____

Mental recall of addition within 8 ◀ Blue Pupil Book Part 1 pages 18 and 19

Number Connections © Rose Griffiths 2005
Harcourt Education Ltd

Spiders and snakes

Name _______________________

Date _______________________

Each spider is made with ten pipecleaners.
How many pipecleaners in each box?

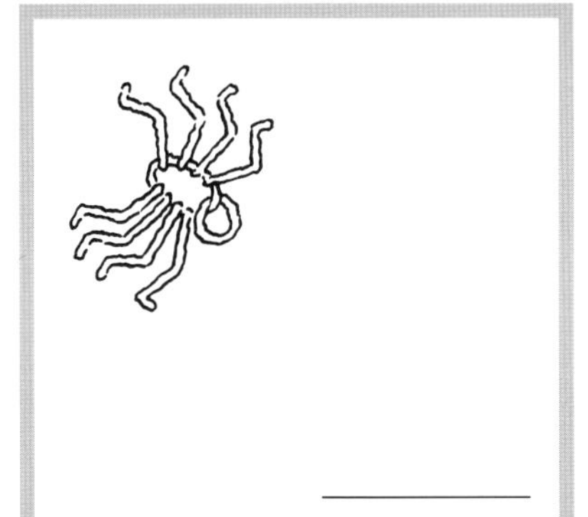

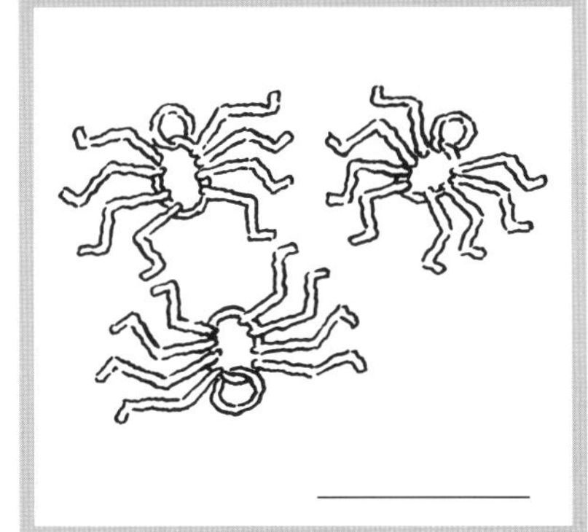

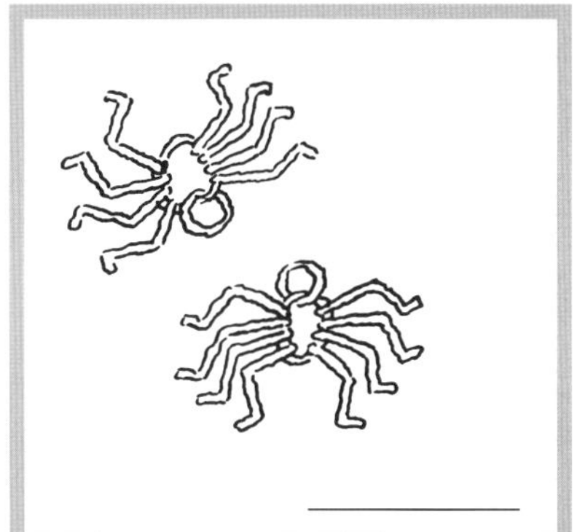

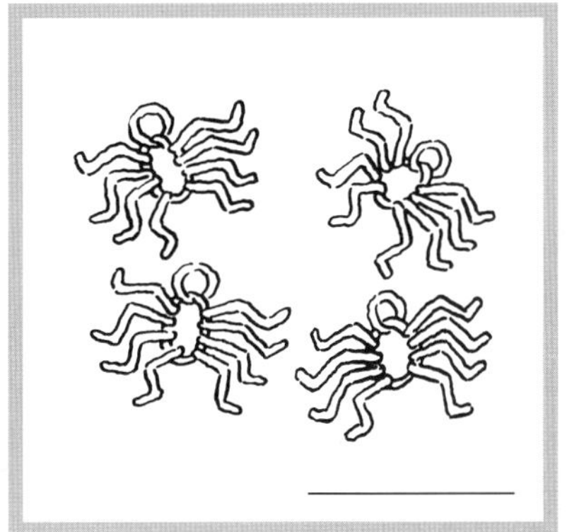

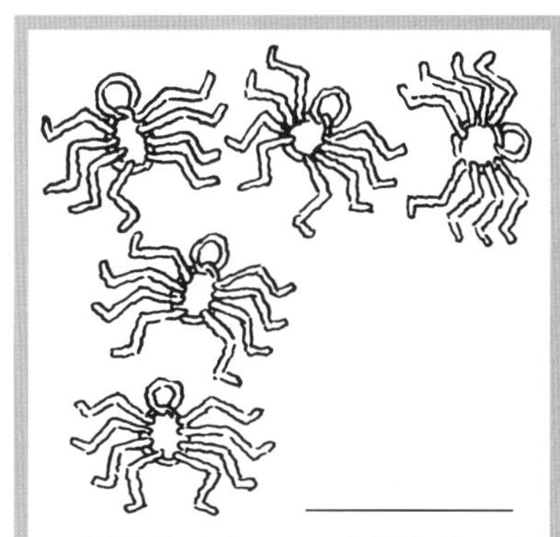

How many pipecleaners?

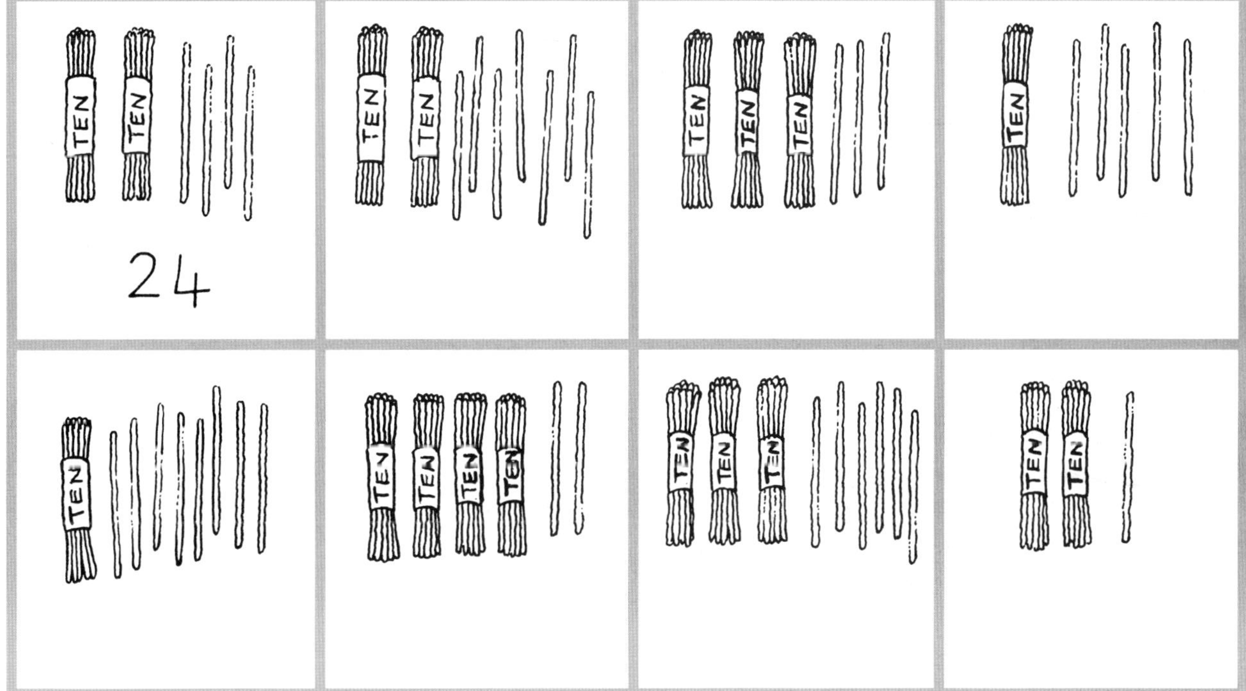

Spiders and snakes

Name ___________________

Date ___________________

Each snake is made with two pipecleaners.
How many pipecleaners in each box?

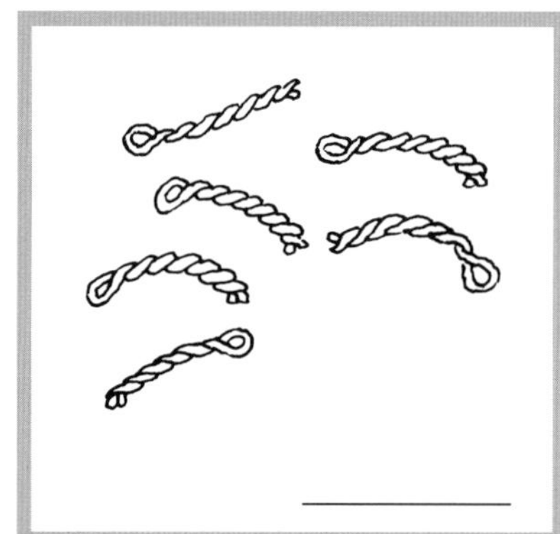

Fill in the chart.

	How many snakes?	How many pipecleaners?

What comes next?

2, 4, 6, 8, ______

20, 22, 24, ______

32, 34, 36, 38, ______

42, 44, 46, ______, ______,

Speedy sums Ⓐ
3 minute test

Name _______________

Date _______________

4 + 2 = ____	5 + 2 = ____	2 + 2 = ____
3 + 4 = ____	1 + 6 = ____	0 + 7 = ____
1 + 3 = ____	2 + 5 = ____	4 + 3 = ____
6 + 1 = ____	0 + 2 = ____	3 + 2 = ____
2 + 3 = ____	4 + 1 = ____	3 + 3 = ____
5 + 0 = ____	2 + 4 = ____	1 + 4 = ____
3 + 1 = ____	1 + 5 = ____	 Score: ____

Mental recall of bonds within 8 ◄ Blue Pupil Book Part 1 pages 22 and 23 *Number Connections* © Rose Griffiths 2005 Harcourt Education Ltd

---

Speedy sums Ⓑ
3 minute test

Name _______________

Date _______________

2 + 5 = ____	4 − 2 = ____	1 + 2 = ____
4 + 2 = ____	7 − 3 = ____	5 − 4 = ____
0 + 6 = ____	5 − 5 = ____	3 + 3 = ____
2 + 3 = ____	6 − 0 = ____	6 − 2 = ____
6 + 1 = ____	7 − 4 = ____	7 + 0 = ____
4 + 3 = ____	3 − 1 = ____	6 − 3 = ____
1 + 3 = ____	7 − 2 = ____	 Score: ____

Mental recall of bonds within 8 ◄ Blue Pupil Book Part 1 pages 22 and 23 ► Copymaster B14 *Number Connections* © Rose Griffiths 2005 Harcourt Education Ltd

Speedy sums C

1   2   3   minute test

Name ___________________

Date ___________________

2 + 6 = ____   7 − 3 = ____   4 + 4 = ____

4 + 2 = ____   6 − 6 = ____   5 − 2 = ____

1 + 1 = ____   8 − 7 = ____   3 + 3 = ____

3 + 5 = ____   4 − 1 = ____   8 − 6 = ____

2 + 3 = ____   8 − 2 = ____   4 + 0 = ____

3 + 4 = ____   7 − 5 = ____   8 − 3 = ____

2 + 5 = ____   7 − 0 = ____    Score: ____

Mental recall of bonds within 8   ◄ Blue Pupil Book Part 1
pages 22 and 23 onwards

Number Connections © Rose Griffiths 2005
Harcourt Education Ltd

Speedy sums D

1   2   3   minute test

Name ___________________

Date ___________________

2 + 4 = ____   7 − 2 = ____   3 + 0 = ____

3 + 3 = ____   6 − 5 = ____   5 − 3 = ____

6 + 2 = ____   8 − 8 = ____   4 + 4 = ____

5 + 3 = ____   5 − 1 = ____   8 − 5 = ____

1 + 1 = ____   4 − 2 = ____   1 + 3 = ____

4 + 3 = ____   7 − 4 = ____   8 − 4 = ____

3 + 2 = ____   7 − 1 = ____    Score: ____

Mental recall of bonds within 8   ◄ Blue Pupil Book Part 1
pages 22 and 23 onwards

Number Connections © Rose Griffiths 2005
Harcourt Education Ltd

T-shirts

Name _______________________

Date _______________________

✂ -

These questions are for _________________________________

How much change? _________

How much change? _________

How much change? _________

How much change? _________

How much change? _________

How much change? _________

T-shirts

Name _______________________

Date _______________________

These questions are for _______________________________

How much did I spend? __________

How much did I spend? __________

How much did I spend? __________

How much did I spend? __________

Bowling

Name ___________________

Date ___________________

Fill in the missing numbers.

| Skittles |

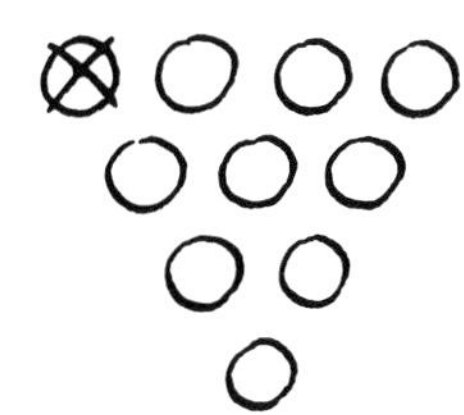

Knocked down: 4
Still standing: 6
____
10

Knocked down:
Still standing: ______

Knocked down:
Still standing: ______

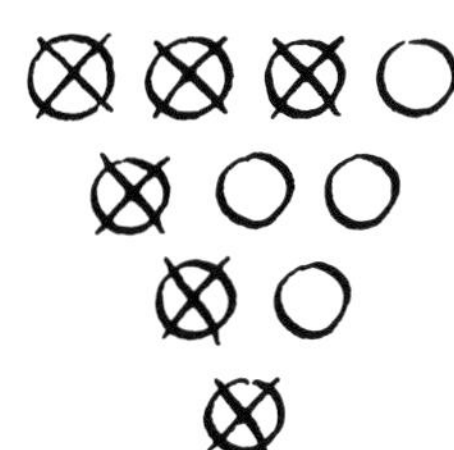

Knocked down:
Still standing: ______
______

Knocked down:
Still standing: ______
______

Knocked down:
Still standing: ______
______

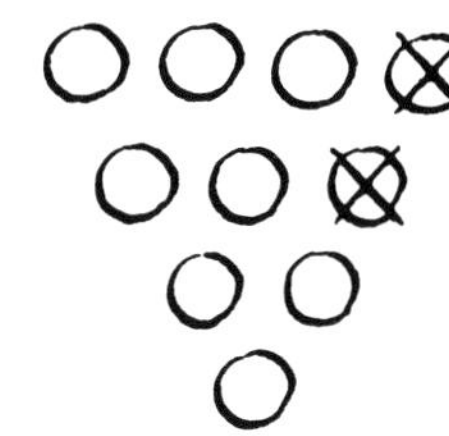

Knocked down:
Still standing: ______
______

Knocked down:
Still standing: ______
______

Knocked down:
Still standing: ______
______

Fives and ones

Name _______________________________

Date _______________________________

Draw the missing 5p coins.

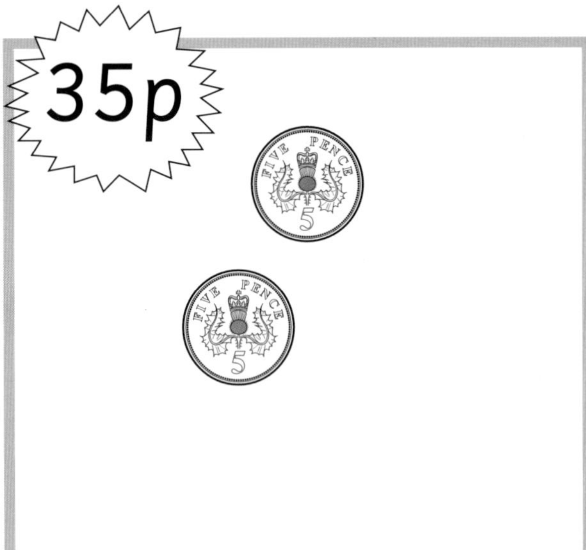

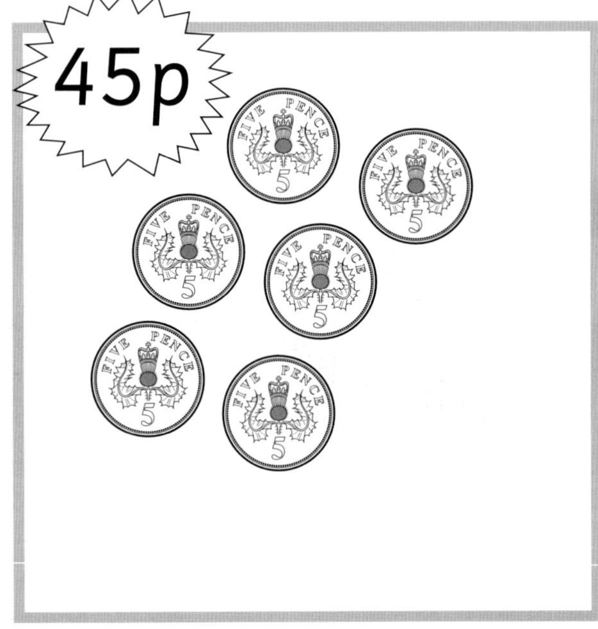

Fives and ones

Name _______________________

Date _______________________

Fill in the missing numbers.
Use coins to help you.

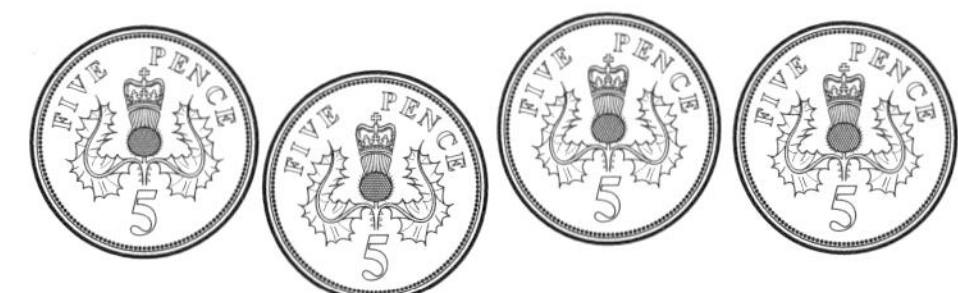

5p + 5p = _______

15p + 5p = _______

25p + 5p = _______

35p + 5p = _______

45p + 5p = _______

10p + 5p = _______

20p + 5p = _______

30p + 5p = _______

40p + 5p = _______

10p + 10p = _______   15p + 15p = _______

20p + 20p = _______   25p + 25p = _______

Pick up bricks

Name _______________________

Date _______________________

Draw the missing bricks.

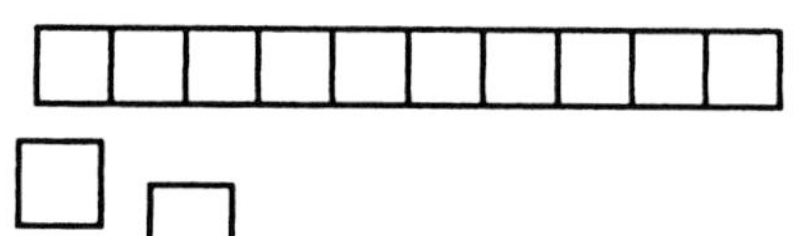

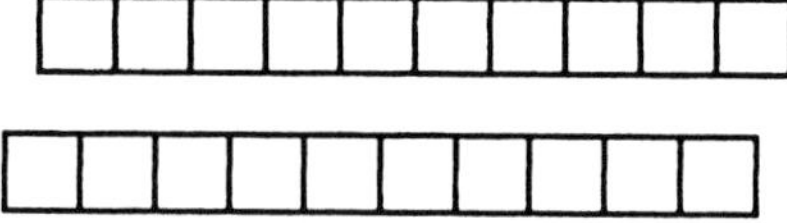

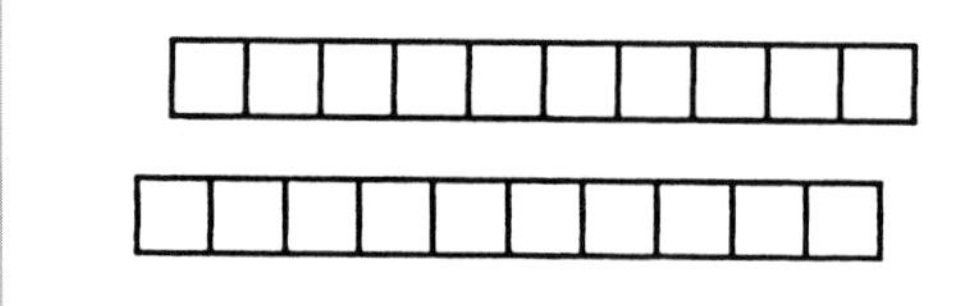

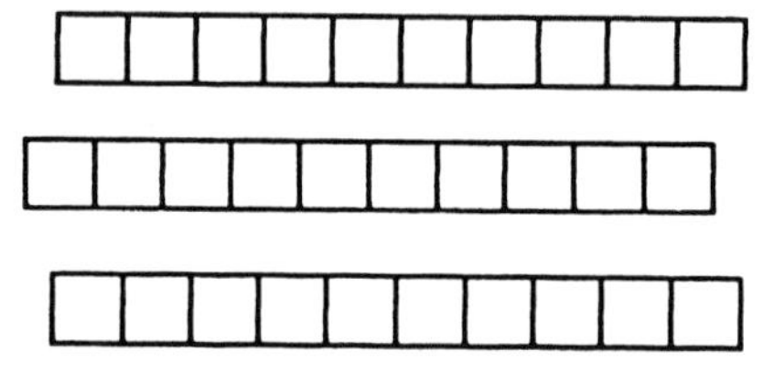

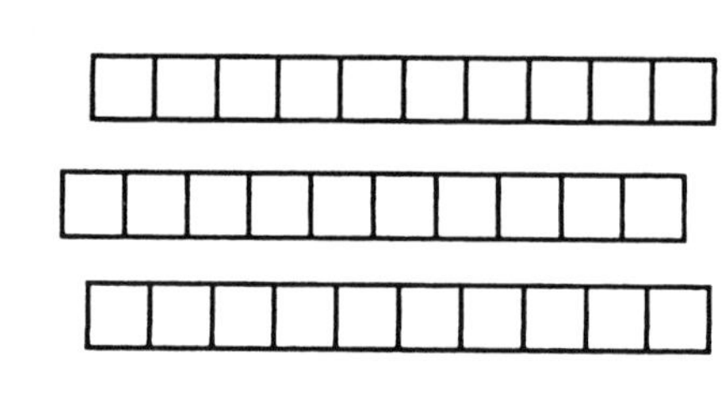

Pick up bricks

Name ________________

Date ________________

Count the bricks.   Do the sums.

10 + 1 = ____

10 + 2 = ____

10 + 3 = ____

10 + 4 = ____

10 + 5 = ____

10 + 6 = ____

10 + 7 = ____

10 + 8 = ____

10 + 9 = ____

10 + 10 = ____

20 − 2 = ____

20 − 4 = ____

20 − 6 = ____

20 − 8 = ____

20 − 10 = ____

Bat and fives

Name _______________________

Date _______________________

Fill in the missing numbers.

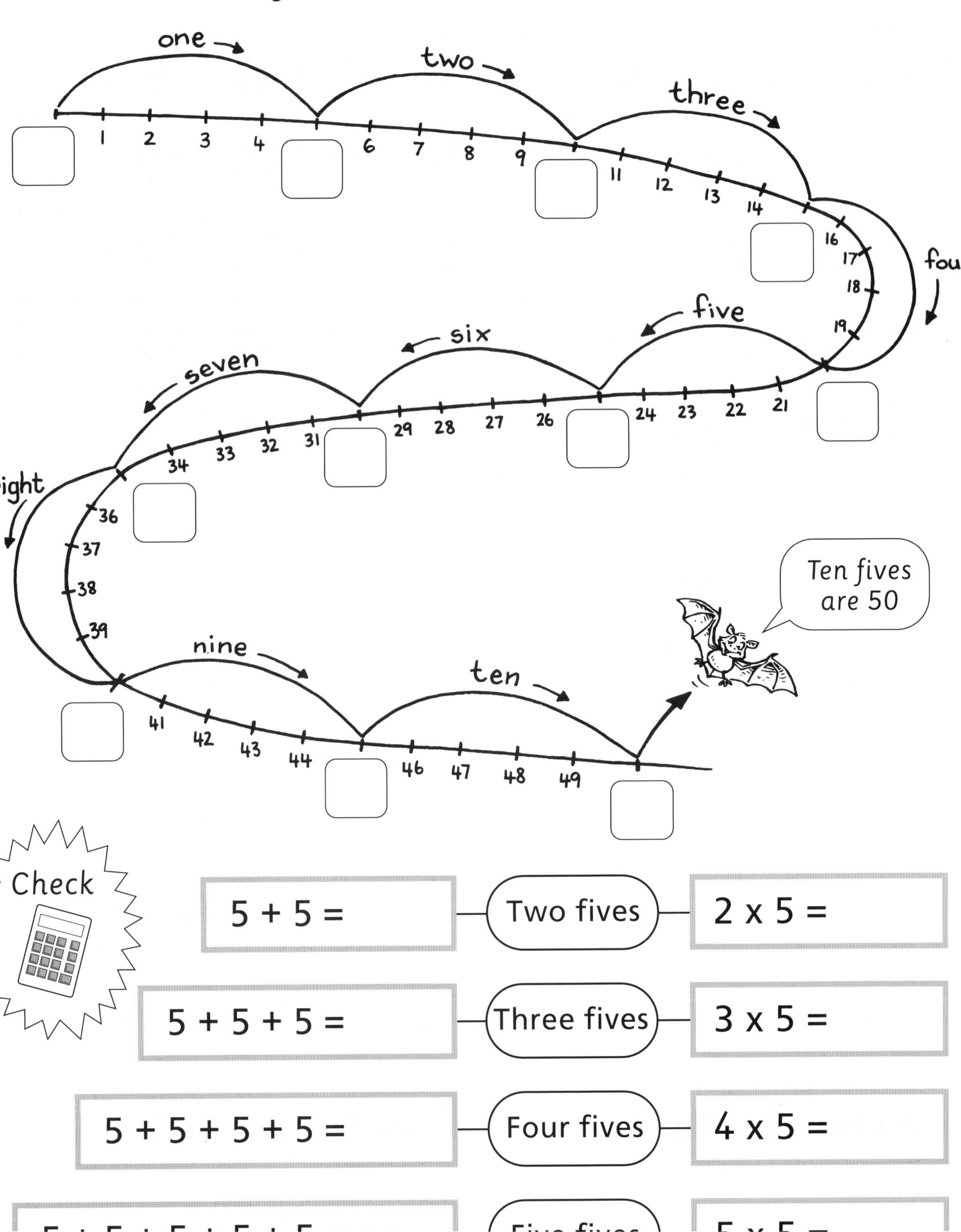

5 + 5 =	Two fives	2 x 5 =
5 + 5 + 5 =	Three fives	3 x 5 =
5 + 5 + 5 + 5 =	Four fives	4 x 5 =
5 + 5 + 5 + 5 + 5 =	Five fives	5 x 5 =

Bat and fives

Name _______________________

Date _______________________

Fill in the missing numbers.

1	2	3	4	5	6	7	8	9	
11	12	13	14		16	17	18	19	
21	22	23	24		26	27	28	29	
31	32	33	34		36	37	38	39	
41	42	43	44		46	47	48	49	

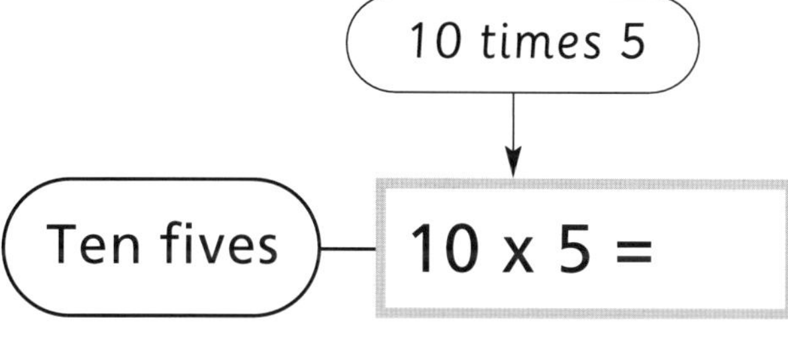
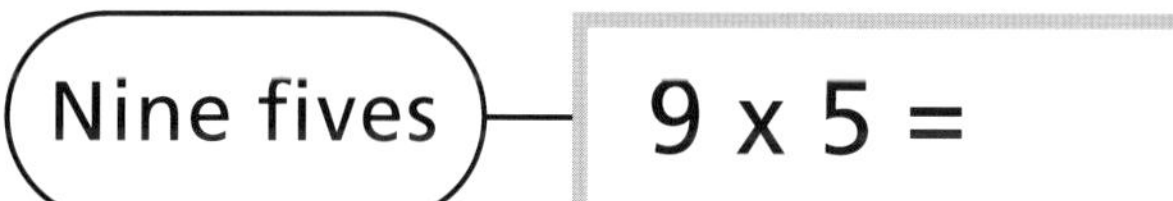

10 times 5

Ten fives	10 x 5 =	Five fives	5 x 5 =
Nine fives	9 x 5 =	Four fives	4 x 5 =
Eight fives	8 x 5 =	Three fives	3 x 5 =
Seven fives	7 x 5 =	Two fives	2 x 5 =
Six fives	6 x 5 =	One five	1 x 5 =
		No fives	0 x 5 =

Card sums

Name ___________________

Date ___________________

Cut out these cards. Use them with Copymaster B25.

Card sums

Name _______________________

Date _______________________

You need the cards made from Copymaster B24.
Find cards to make each sum. Stick them on.

> 2 cards which
> add up to 20.

> 2 cards which
> add up to 17.

> 2 cards which
> add up to 10.

> 2 cards which
> add up to 14.

Work with a friend.
Make up some sums like these for each other.
Use the rest of your cards.

Sums which make 8

sheet 1 of 3

Print on card if possible. Reusable.
Cut out the instructions card, 3 sum cards and 34 number cards. Store in a clear zip-top wallet or in an envelope.

≈ Sums which make 8 ≈

- **Before you start**
  Take a sum card each.
  Shuffle the number cards.
  Put them in a pile on the table, face down.

- **How to play**
  Can you make 4 different sums which make 8?

- **Keep going until you have all filled your cards.**
  Who finished first?

◀ Blue Pupil Book Part 1; **Mental recall of addition within 8**

Number Connections © Rose Griffiths 2005
Harcourt Education Ltd

Sums which make 8

sheet 2 of 3

Print on card if possible. Reusable.

0	1	1	2	2
0	1	1	2	2
3	3	4	4	4
3	3	4	4	4
5	5	6 (six)	6 (six)	7
5	5	6 (six)	6 (six)	7
8	8		7	7

Sums which make 8

sheet 3 of 3

Print 3 copies, on card if possible. Reusable.

Sums which make 8

☐ + ☐ = 8

☐ + ☐ = 8

☐ + ☐ = 8

☐ + ☐ = 8

Fifty pence game

sheet 1 of 2

T

GP

Print on card if possible. Reusable.
Cut out the instructions card and 16 playing cards. Store in a clear zip-top wallet or in an envelope.

- **Before you start**
  Shuffle the cards.
  Spread them out on the table, face down.

- **How to play**

- **Keep going until no one can make any more fifty pences.**

◀ Blue Pupil Book Part 1; **Counting in 5s to 50**

Number Connections © Rose Griffiths 2005
Harcourt Education Ltd

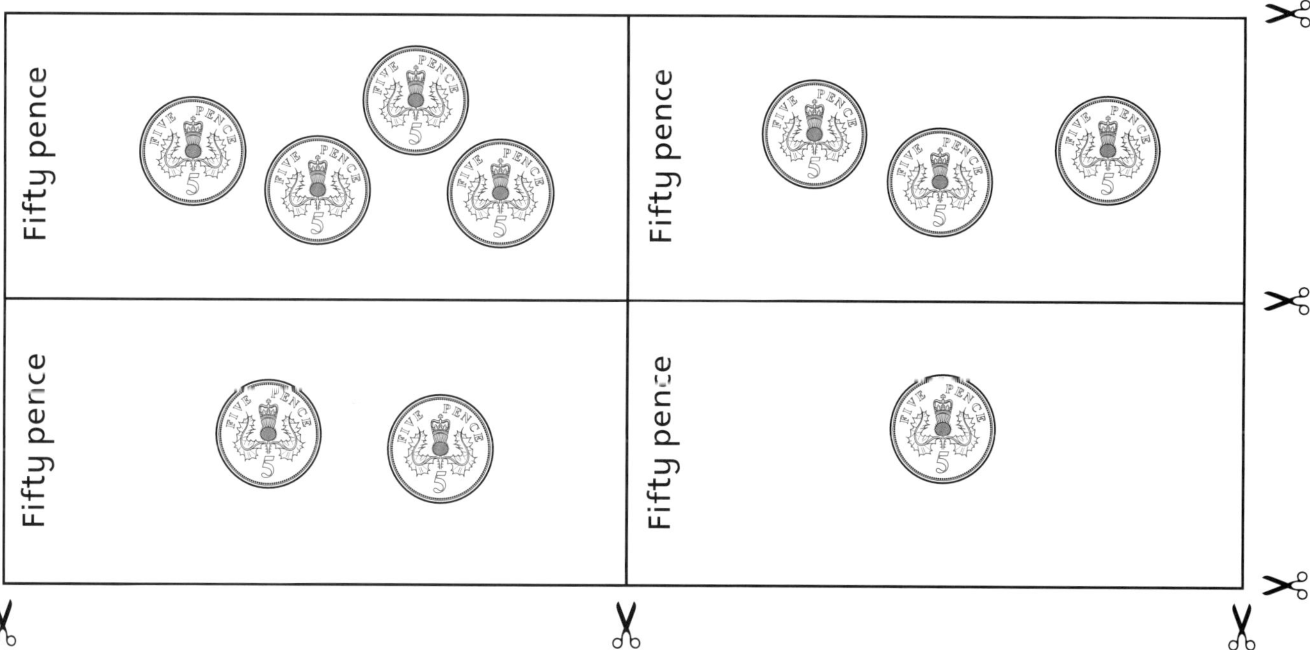

Fifty pence game

sheet 2 of 2

Print on card if possible. Reusable.

Fifty pence	Fifty pence
Fifty pence	Fifty pence
Fifty pence	Fifty pence
Fifty pence	Fifty pence
Fifty pence	Fifty pence
Fifty pence	Fifty pence

Coins in a jar

Name ________________________

Date ________________________

How much money?

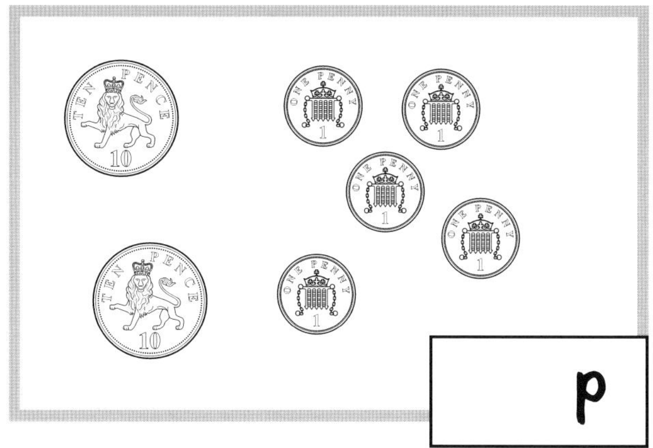

p

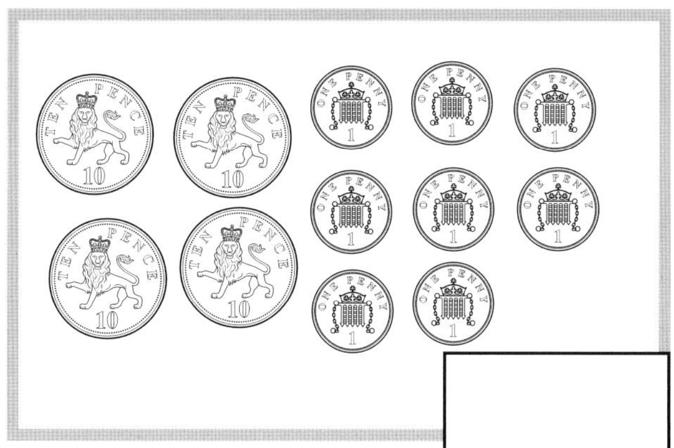

Counting to 60 ◄ Blue Pupil Book Part 2 pages 38 and 39

Number Connections © Rose Griffiths 2005
Harcourt Education Ltd

Sums in words

Name _______________________

Date _______________________

11

eleven
elev___
el______
e________
eleven

12

twelve
twel___
tw______
t________
twelve

13

thirteen
thir______
th________
____teen
th________
____
thirteen

14

fourteen
four______
f__________
____teen
f__________
____
fourteen

15

fifteen
fif______
f________
f___teen
f________
____
fifteen

Mixed up numbers! Sort them out.

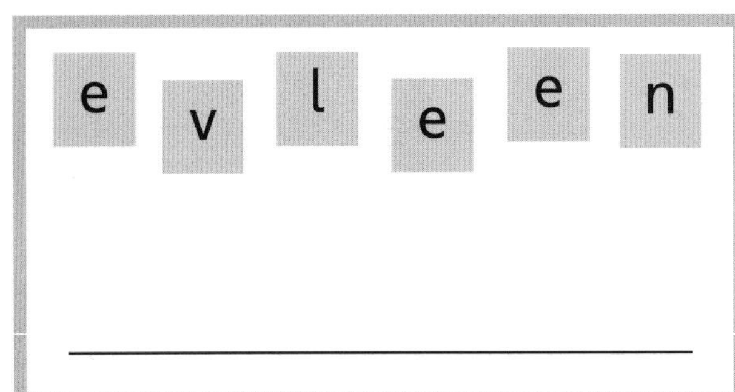

____

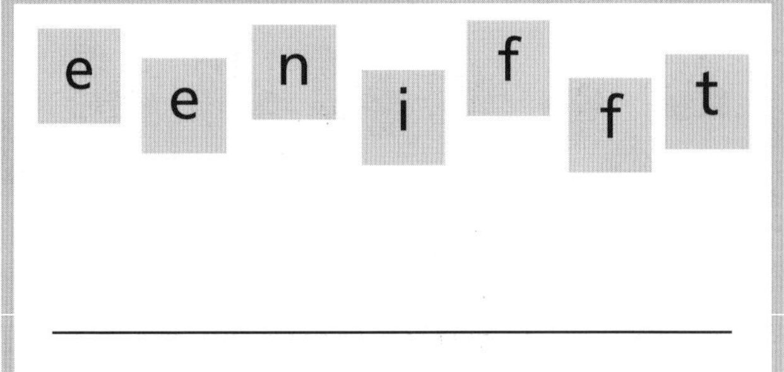

____

Sums in words

Name ______________________

Date ______________________

16

s i x t e e n
s i x _______
s _________
____
s i x t e e n

17

s e v e n t e e n
s e v e n _______
s e v _________
s _________
s e v e n t e e n

18

e i g h t e e n
e i g h t _______
e i _________
_______ t e e n
e _________
____
e i g h t e e n

19

n i n e t e e n
n i n e _______
n _________
_______ t e e n
n _________
____
n i n e t e e n

20

t w e n t y
t w e n _______
t w _______
_______ t y
t _________
____
t w e n t y

Mixed up numbers! Sort them out.

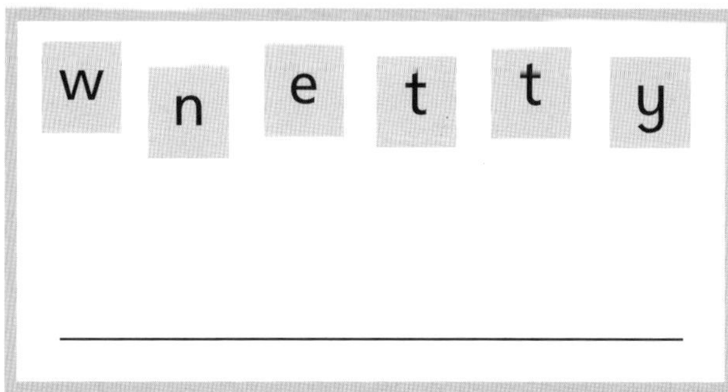

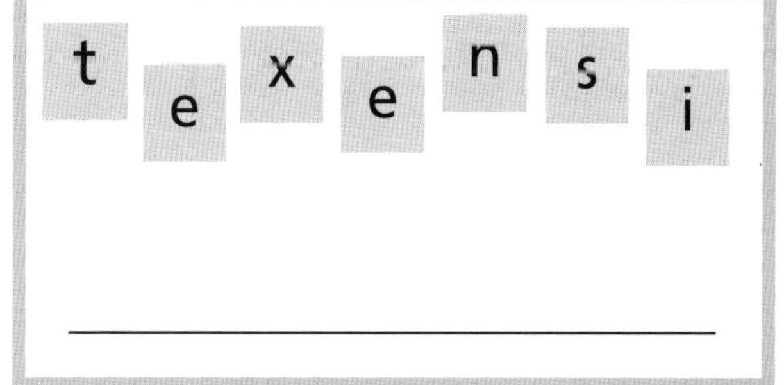

Tens and ones

Name _______________________

Date _______________________

Draw it.   Then add one.

| 18 |

18 + 1 = _____

| 49 |

49 + 1 = _____

| 23 |

23 + 1 = _____

| 31 |

31 + 1 = _____

| 42 |

42 + 1 = _____

| 30 |

30 + 1 = _____

Tens and ones

Name _______________________

Date _______________________

Draw it.

Then add ten.

18

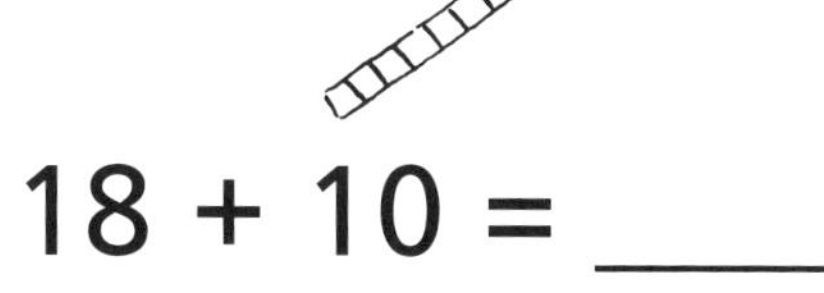

18 + 10 = _____

49

49 + 10 = _____

23

23 + 10 = _____

31

31 + 10 = _____

42

42 + 10 = _____

30

30 + 10 = _____

Two times table

Name _______________________

Date _______________________

Cut out the eleven tables facts.
Fold along the dotted line and glue flat.

		5 × 2	10
0 × 2	0	6 × 2	12
1 × 2	2	7 × 2	14
2 × 2	4	8 × 2	16
3 × 2	6	9 × 2	18
4 × 2	8	10 × 2	20

Two times table

Name _______________________

Date _______________________

Fill in the missing numbers.
Check with a calculator.

5 × 2 = ☐
2 × 5 = ☐

8 × 2 = ☐
2 × 8 = ☐

3 × 2 = ☐
2 × 3 = ☐

10 × 2 = ☐
2 × 10 = ☐

10 × 2 = ☐
☐ × 2 = 18
8 × 2 = ☐
7 × ☐ = 14
☐ × 2 = 12
5 × 2 = ☐
4 × 2 = ☐
☐ × 2 = 6
2 × ☐ = 4
1 × 2 = ☐
0 × 2 = ☐

____

____

Hours and half hours

Name ________________________

Date ________________________

Make your clock show each time. Then draw it.

One o'clock	Half past one	Two o'clock
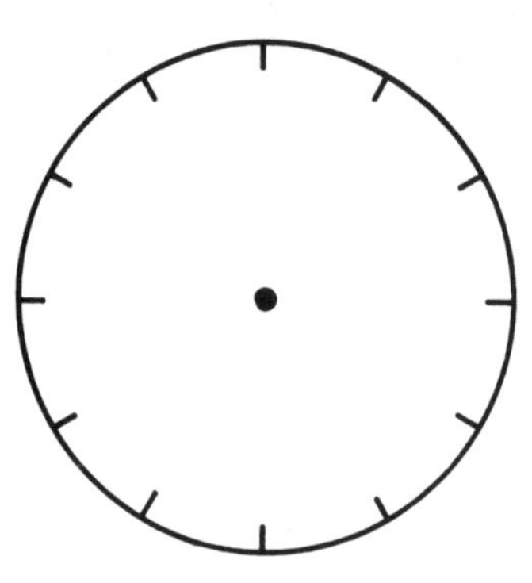	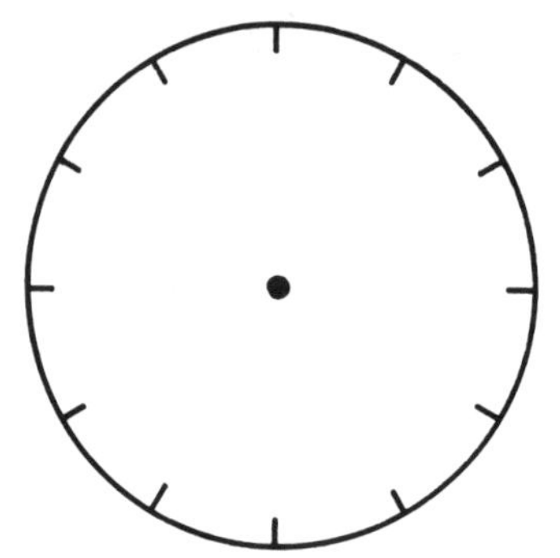	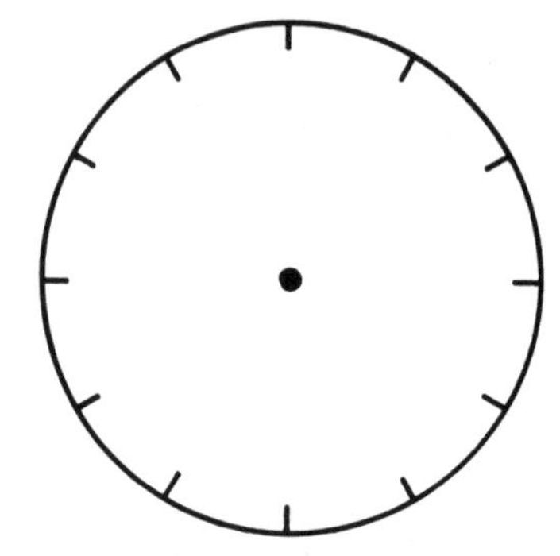
Half past two	Three o'clock	Half past three
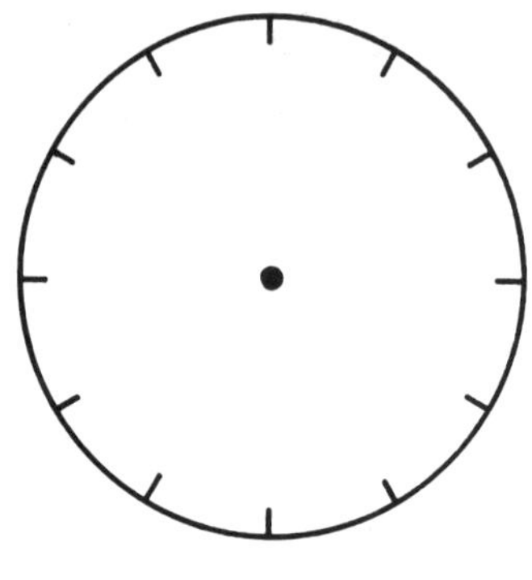	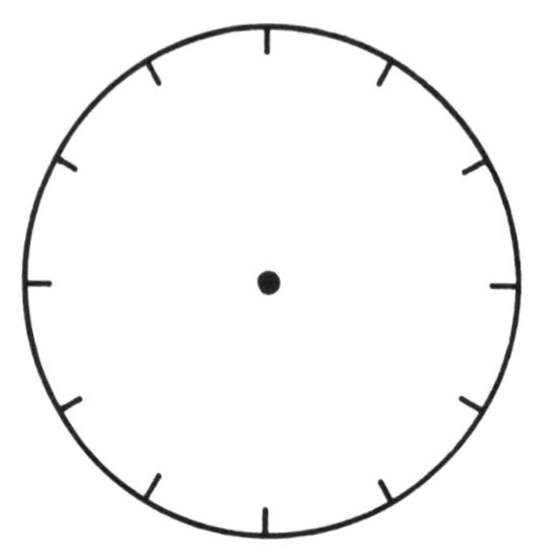	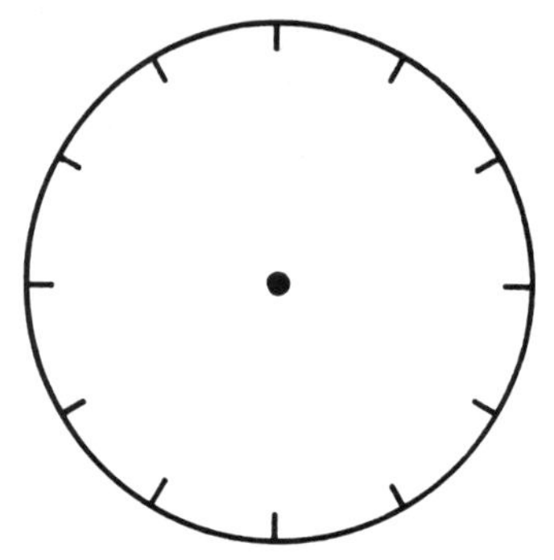
Four o'clock	Half past four	Five o'clock
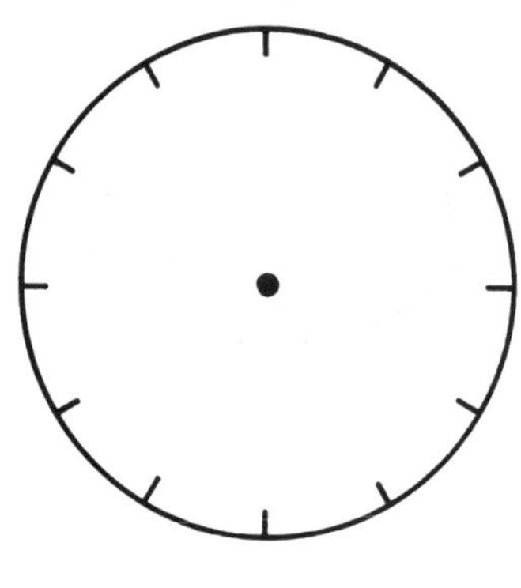	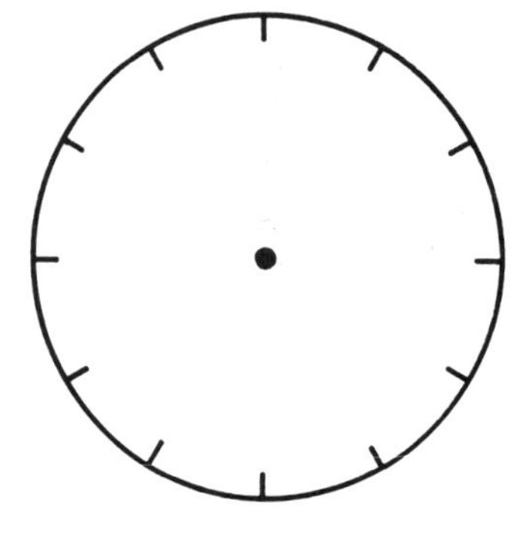	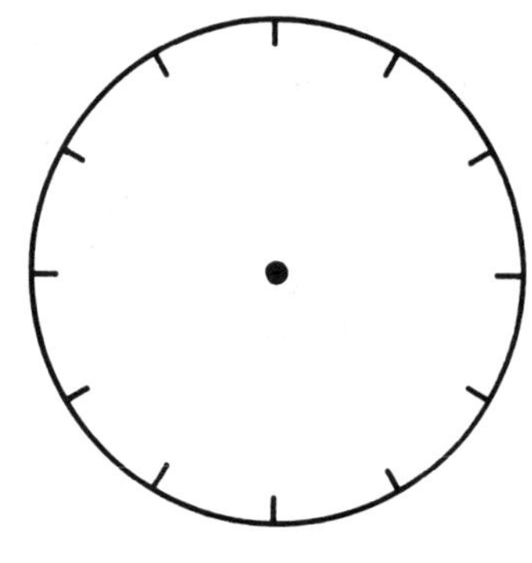

Hours and half hours

Name _______________________

Date _______________________

4 o'clock

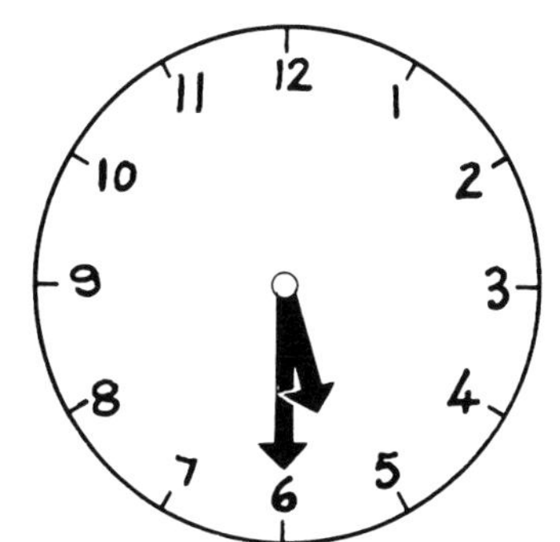

Half past ____

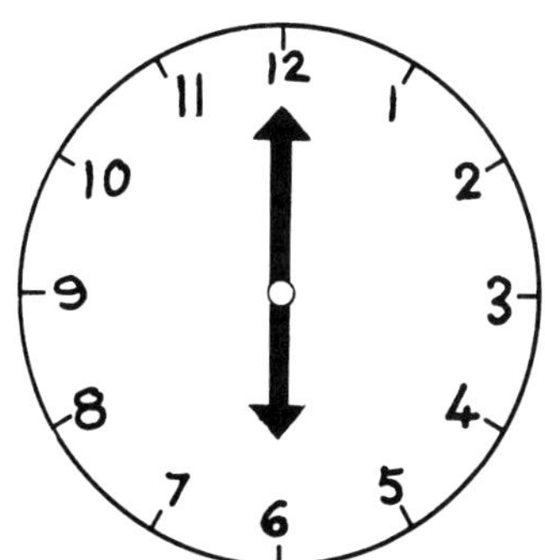

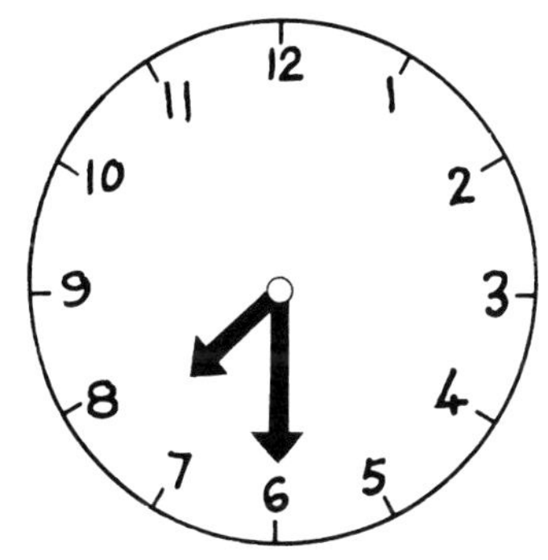

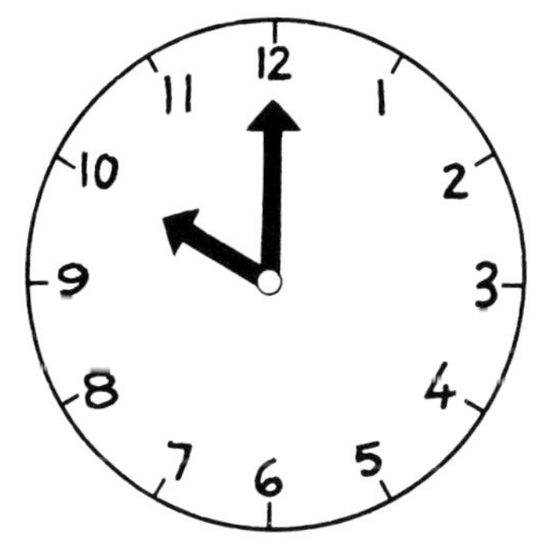

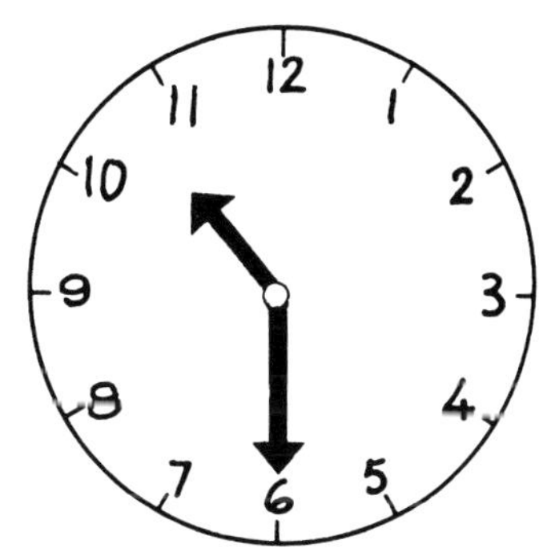

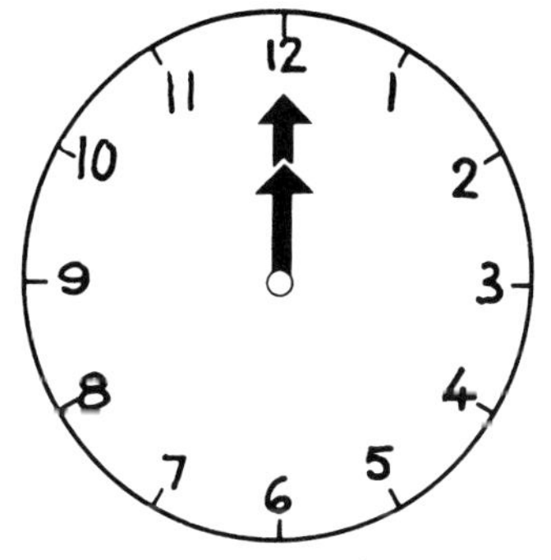

Nine counters

Name _______________________

Date _______________________

3 red, **6** blue.

| 3 | + | 6 | = | 9 |

___ red, ___ blue.

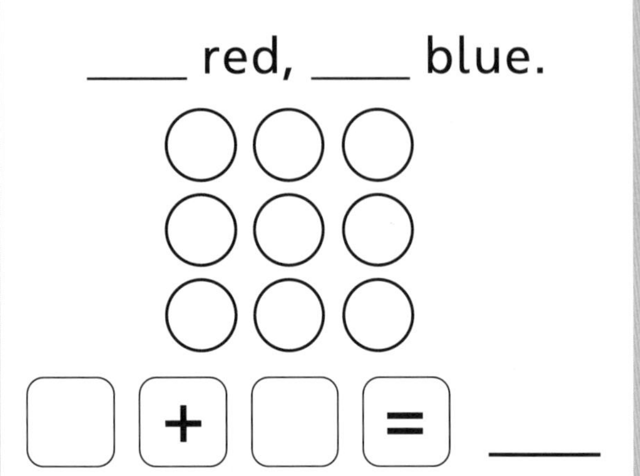

☐ + ☐ = ___

___ red, ___ blue.

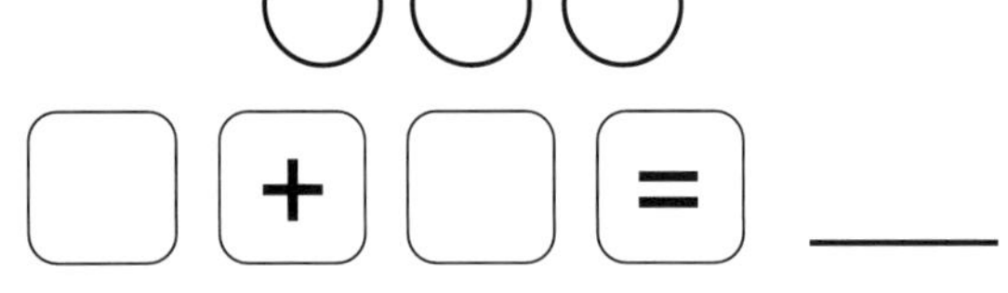

☐ + ☐ = ___

___ red, ___ blue.

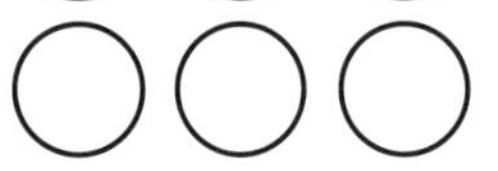

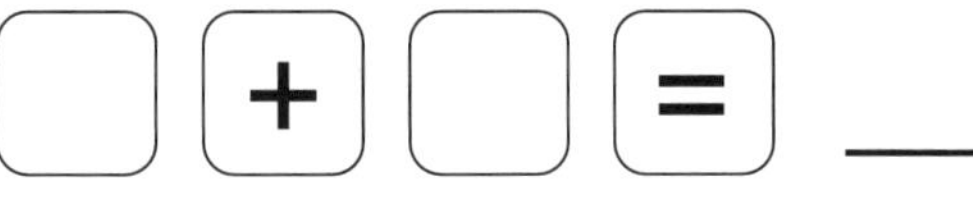

☐ + ☐ = ___

___ red, ___ blue.

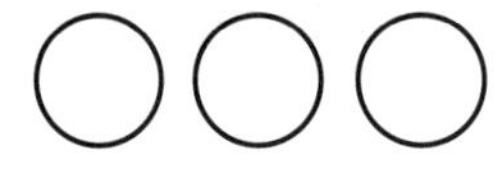

☐ + ☐ = ___

___ red, ___ blue.

☐ + ☐ = ___

Nine counters

Name ______________________

Date ______________________

2 + ☐ = 9
4 + ☐ = 9
6 + ☐ = 9
3 + ☐ = 9
8 + ☐ = 9
5 + ☐ = 9
1 + ☐ = 9
7 + ☐ = 9

☐ + 0 = 9
☐ + 5 = 9
☐ + 3 = 9
☐ + 6 = 9
☐ + 2 = 9
☐ + 4 = 9
☐ + 8 = 9
☐ + 7 = 9

4 + ☐ = 9
☐ + 4 = 9

5 + ☐ = 9
☐ + 5 = 9

6 + ☐ = 9
☐ + 6 = 9

3 + ☐ = 9
☐ + 3 = 9

Fives and tens

Name _______________________

Date _______________________

Fill in the four missing numbers.

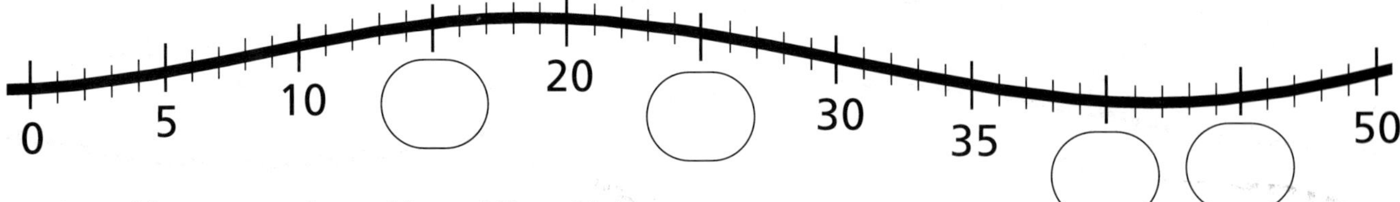

Use the number line like this:

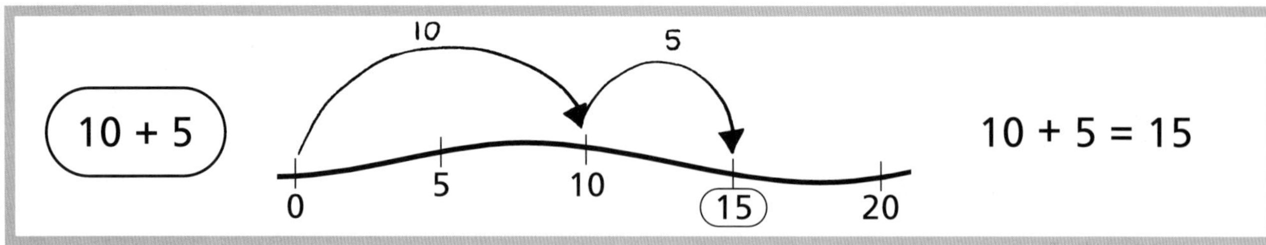

$10 + 5 = 15$

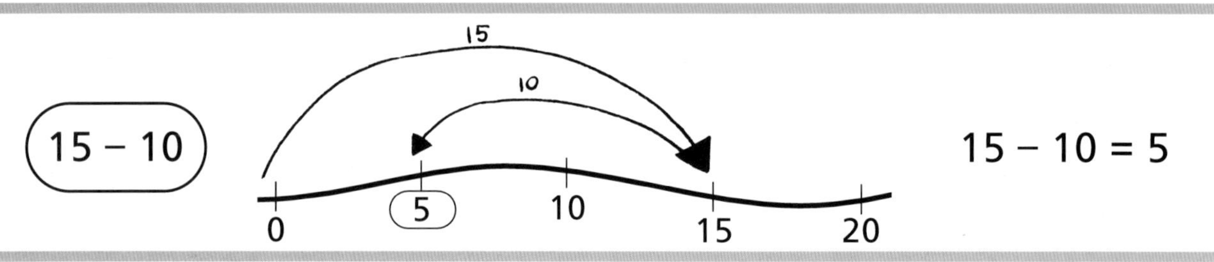

$15 - 10 = 5$

$10 + 10 = $ _____	$20 - 15 = $ _____
$10 + 15 = $ _____	$30 - 10 = $ _____
$20 + 15 = $ _____	$45 - 5 = $ _____
$15 + 20 = $ _____	$50 - 10 = $ _____
$15 + 25 = $ _____	$45 - 20 = $ _____
$25 + 25 = $ _____	$45 - 15 = $ _____
$25 + 10 = $ _____	$30 - 15 = $ _____
$45 + 5 = $ _____	$25 - 20 = $ _____

Fives and tens

Name _______________________

Date _______________________

Use the number line then check with a calculator.

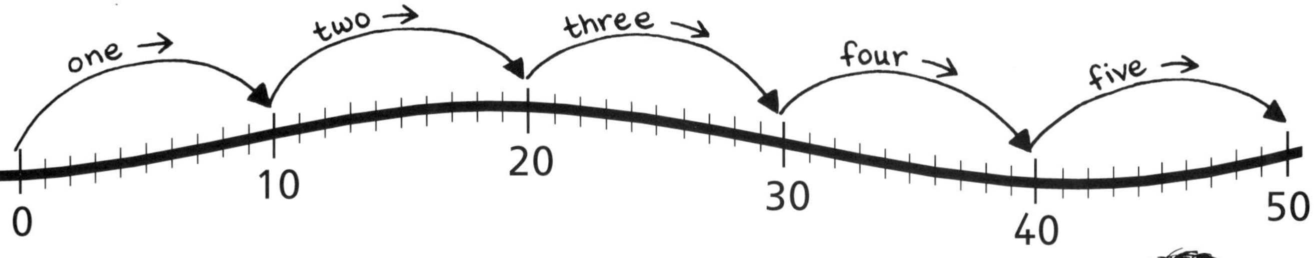

No tens — 0 × 10 =

One ten — 1 × 10 =

Two tens — 2 × 10 =

Three tens — 3 × 10 =

Four tens — 4 × 10 =

Five tens — 5 × 10 =

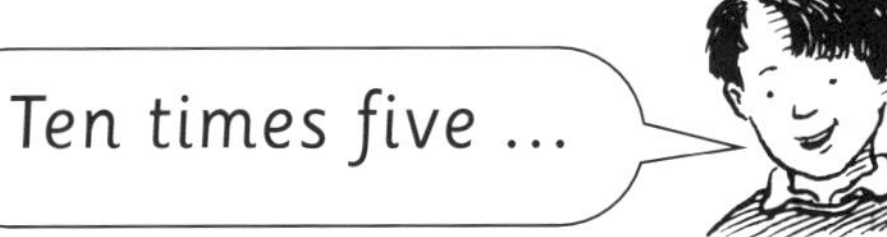

10 × 5 = ____

10 × 4 = ____

10 × 3 = ____

10 × 2 = ____

10 × 1 = ____

10 × 0 = ____

2 × 10 = ____

4 × 10 = ____

3 × 10 = ____

0 × 10 = ____

5 × 10 = ____

10 × 1 = ____

10 × 3 = ____

10 × 5 = ____

10 × 4 = ____

10 × 2 = ____

Speedy sums Ⓔ

1   2   3   minute test

Name _______________

Date _______________

$5 + 2 =$ ____	$8 - 3 =$ ____	$3 + 3 =$ ____
$2 + 3 =$ ____	$7 - 6 =$ ____	$5 - 3 =$ ____
$4 + 4 =$ ____	$3 - 3 =$ ____	$4 + 5 =$ ____
$5 + 0 =$ ____	$9 - 5 =$ ____	$8 - 2 =$ ____
$1 + 8 =$ ____	$7 - 2 =$ ____	$3 + 4 =$ ____
$6 + 3 =$ ____	$2 - 1 =$ ____	$6 - 4 =$ ____
$2 + 4 =$ ____	$9 - 2 =$ ____	 Score: ____

Speedy sums Ⓕ

1   2   3   minute test

Name _______________

Date _______________

$6 + 3 =$ ____	$9 - 8 =$ ____	$4 + 4 =$ ____
$4 + 3 =$ ____	$8 - 5 =$ ____	$9 - 4 =$ ____
$3 + 0 =$ ____	$7 - 3 =$ ____	$6 + 2 =$ ____
$2 + 2 =$ ____	$9 - 7 =$ ____	$8 - 3 =$ ____
$5 + 4 =$ ____	$6 - 2 =$ ____	$1 + 5 =$ ____
$2 + 5 =$ ____	$9 - 3 =$ ____	$6 - 5 =$ ____
$4 + 1 =$ ____	$4 - 4 =$ ____	 Score: ____

Boxes

Name _______________________

Date _______________________

Shaun and Emma kept playing 'Boxes'.

5th game

Shaun got 15, and Emma got _____ .

6th game

Shaun got 10, and Emma got _____ .

7th game

Shaun got 5, and Emma got _____ .

8th game

Shaun got 0, and Emma got _____ .

Fill in the missing numbers.

$18 + \boxed{} = 20$ $17 + \boxed{} = 20$

$13 + \boxed{} = 20$ $14 + \boxed{} = 20$

$16 + \boxed{} = 20$ $9 + \boxed{} = 20$

$7 + \boxed{} = 20$ $12 + \boxed{} = 20$

Boxes

Name _______________________

Date _______________________

Play 'Boxes' with a friend. Add up your scores.

1st game

Name	Score
Total	

2nd game

Name	Score
Total	

3rd game

Name	Score
Total	

4th game

Name	Score
Total	

Teen numbers

Name ______________________

Date ______________________

Write the number and the word.

15   fifteen

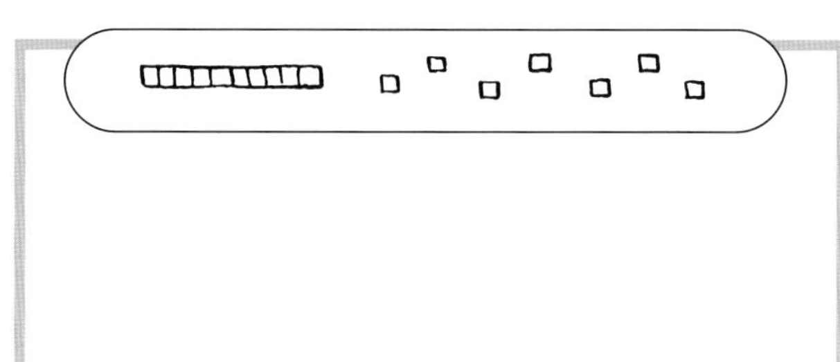

Use tens and ones. Draw them.

16 + 3 =

7 + 11 =

14 + 2 =

10 + 3 =

Teen numbers

Name _______________________

Date _______________________

B48

14 + 5

14 + 5 =

12 + 3

12 + 3 =

6 + 11

6 + 11 =

10 + 7

10 + 7 =

Draw tens and ones.

13 + 4

13 + 4 =

5 + 12

5 + 12 =

8 + 10

8 + 10 =

15 + 3

15 + 3 =

Five times table

Name _______________________

Date _______________________

Cut out the eleven tables facts.
Fold along the dotted line and glue flat.

		5×5	25
0×5	0	6×5	30
1×5	5	7×5	35
2×5	10	8×5	40
3×5	15	9×5	45
4×5	20	10×5	50

Five times table

Name _______________________

Date _______________________

Fill in the missing numbers.
Check with a calculator.

$3 \times 5 = \boxed{}$

$5 \times 3 = \boxed{}$

$10 \times 5 = \boxed{}$

$5 \times 10 = \boxed{}$

$2 \times 5 = \boxed{}$

$5 \times 2 = \boxed{}$

$7 \times 5 = \boxed{}$

$5 \times 7 = \boxed{}$

$10 \times 5 = \boxed{}$

$\boxed{} \times 5 = 45$

$8 \times 5 = \boxed{}$

$7 \times \boxed{} = 35$

$\boxed{} \times 5 = 30$

$\boxed{} \times 5 = 25$

$4 \times 5 = \boxed{}$

$\boxed{} \times 5 = 15$

$\boxed{} \times 5 = 10$

$1 \times 5 = \boxed{}$

$0 \times 5 = \boxed{}$

____

____

More teen numbers

Name _______________________

Date _______________________

16 − 3

16 − 3 =

18 − 10

18 − 10 =

15 − 4

15 − 4 =

13 − 3

13 − 3 =

Cross out tens and ones.

14 − 10

14 − 10 =

17 − 11

17 − 11 =

19 − 8

19 − 8 =

16 − 6

16 − 6 =

Photos

Name _______________________

Date _______________________

How many photos of the rabbit? _____________

How many photos of my friends? _____________

Total: _____________

How many photos of the dog? _____________

How many photos of my friends? _____________

Total: _____________

Photos

Name _______________________

Date _______________________

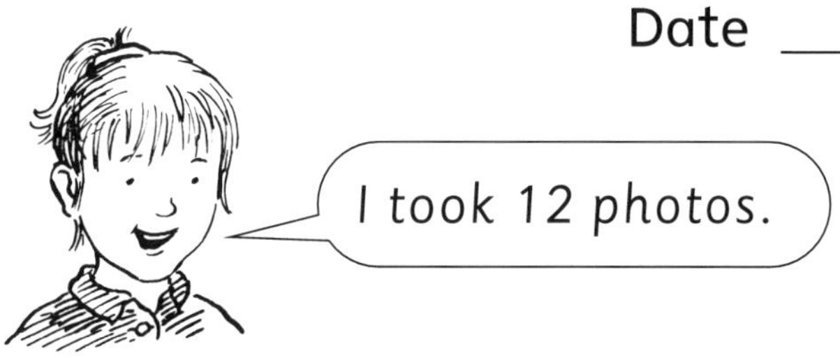

Write or draw to show what was on each photo.
Use these clues:

- The first three photos are my cat. 
- The 4th, 5th and 6th photos are my friends.
- The 11th and 12th photos are my rabbit. 
- There are seven photos of my cat.

1	2	3	4
5	6	7	8
9	10	11	12

What is on the 3rd photo? _______________________

What is on the 8th photo? _______________________

What is on the last photo? _______________________

Hopping frogs

Name _______________________

Date _______________________

Frogs in the dish	Draw the missing frogs.	Score
0		0
1		3
2		
3		
4		
5		
6		

Hopping frogs

Name _______________________

Date _______________________

How many counters?

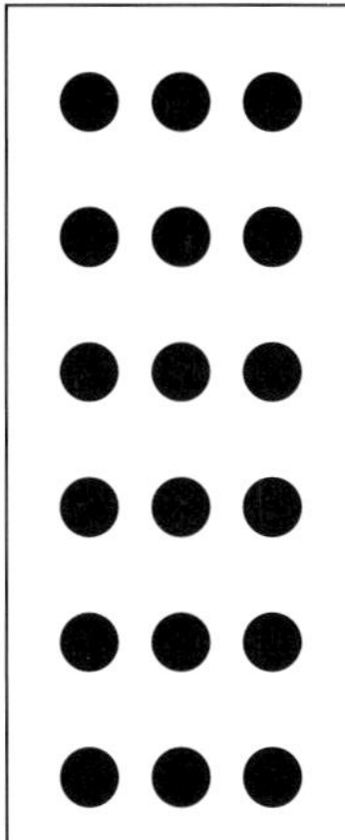 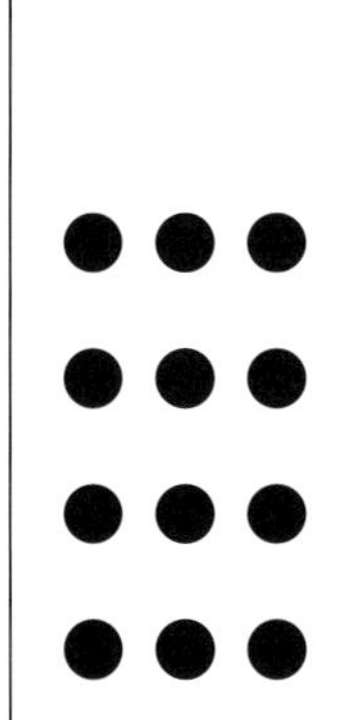 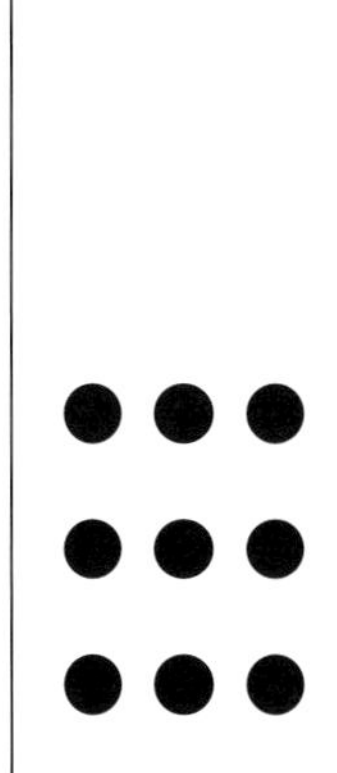 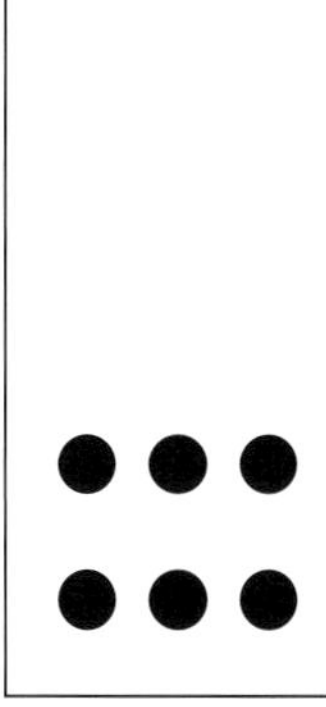 

18 _____   _____   _____   _____   _____   _____

$0 + 3 =$ _____

$3 + 3 =$ _____

$6 + 3 =$ _____

$9 + 3 =$ _____

$12 + 3 =$ _____

$15 + 3 =$ _____

$18 - 3 =$ _____

$15 - 3 =$ _____

$12 - 3 =$ _____

$9 - 3 =$ _____

$6 - 3 =$ _____

$3 - 3 =$ _____

$3 + 12 =$ _____

$15 - 3 =$ _____

$12 + 3 =$ _____

$15 + 3 =$ _____

$18 - 3 =$ _____

$3 + 15 =$ _____

Sixty pence

sheet 1 of 2

Print on card if possible. Reusable.
Colour in and cover with clear plastic if wished, then cut out the two pieces of the board.
Use sticky tape to fasten them together.

≈Sixty pence≈

A game for 1, 2 or 3 people.

- **Before you start**
You need a dice, a counter each,
up to 18 ten pence coins and
30 pennies.

Put your counter on the arrow.

◄ Blue Pupil Book Part 2;
Counting to 60

Sixty pence

sheet 2 of 2

T

GP

Print on card if possible. Reusable.
Store in a clear zip-top wallet or in an envelope.
If possible, include a dice, 3 counters, 18 plastic or card ten pence coins, and 30 pennies.

Number Connections © Rose Griffiths 2005
Harcourt Education Ltd

Tens and ones game

sheet 1 of 2

T

GP

Print on card if possible. Reusable.
Cut out the instructions card, playing board and ten number cards.
Store in a clear zip-top wallet or in an envelope.

≈ Tens and ones ≈

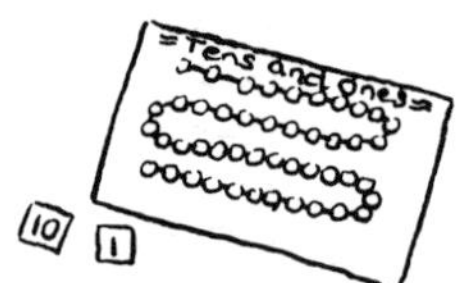

- **Before you start**
  You need a counter each. Put your counter on 0.
  Shuffle the number cards. Put them in a pile, face down.

- **How to play**

- **The first counter to get to 60 is the winner.**

◀ Blue Pupil Book Part 2; **Counting in 10s and 1s to 60**

Number Connections © Rose Griffiths 2005
Harcourt Education Ltd

Tens and ones	Tens and ones	Tens and ones	Tens and ones	Tens and ones
10	10	10	10	10
Tens and ones	Tens and ones	Tens and ones	Tens and ones	Tens and ones
1	1	1	1	1

"

Tens and ones game

sheet 2 of 2

Print on card if possible. Reusable.

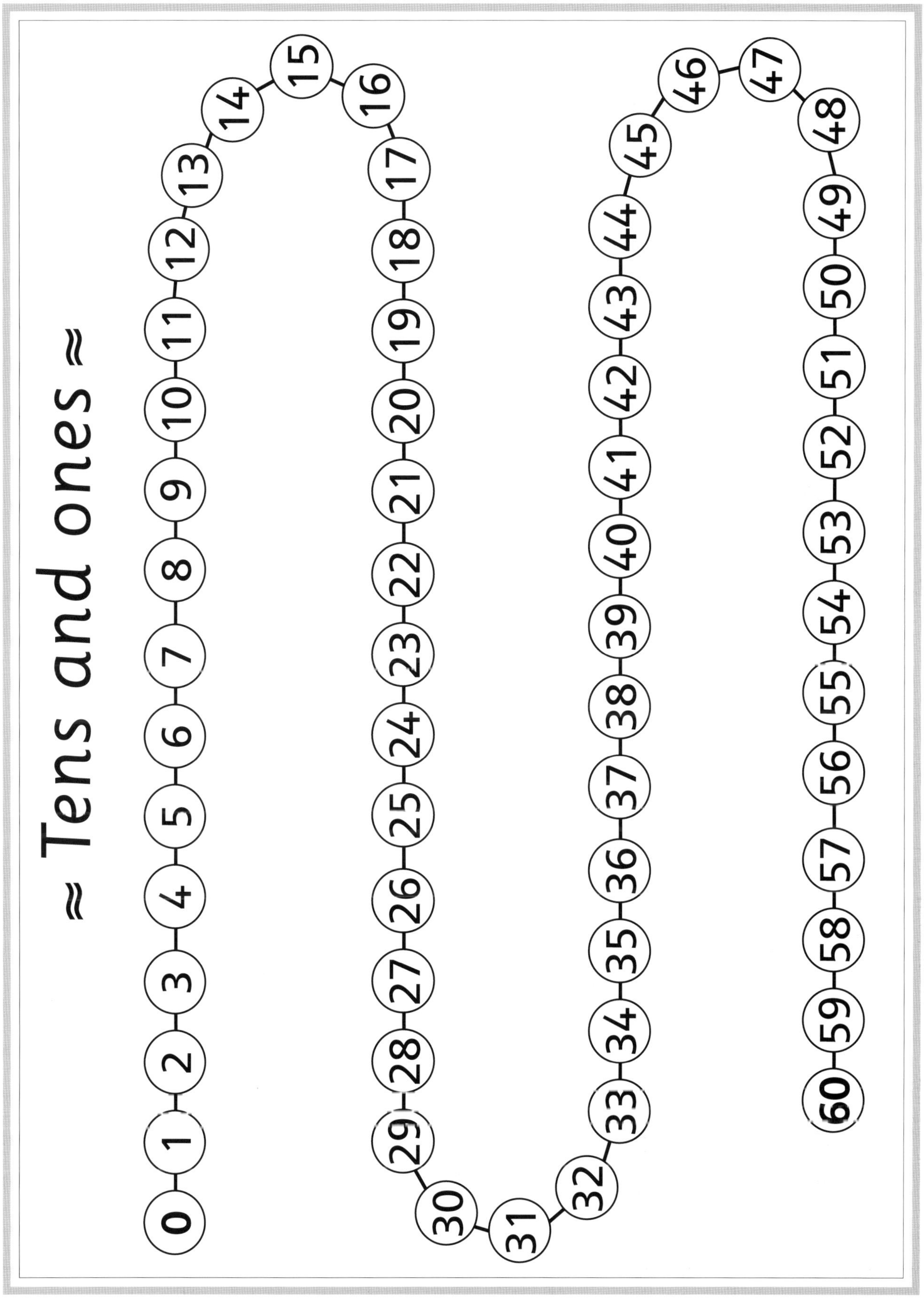

≈ **Tens and ones** ≈

0 1 2 3 4 5 6 7 8 9 10 11 12 13 14 15 16 17 18 19 20 21 22 23 24 25 26 27 28 29 30 31 32 33 34 35 36 37 38 39 40 41 42 43 44 45 46 47 48 49 50 51 52 53 54 55 56 57 58 59 60

Number Connections © Rose Griffiths 2005
Harcourt Education Ltd

Make 9 game

sheet 1 of 2

Print on card if possible. Reusable.
Cut out the instructions card and 20 playing cards.
Store in a clear zip-top wallet or in an envelope.

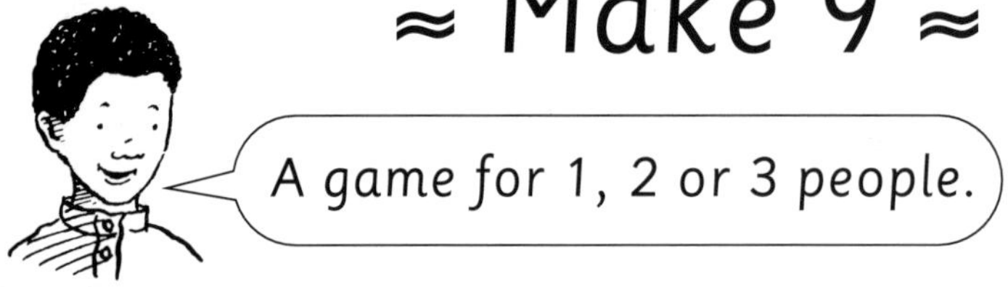

- **Before you start**
  Shuffle the cards.
  Spread them out on the table, face down.

- **How to play**

- **Keep going until all the cards have gone.**

◄ Blue Pupil Book Part 2; **Addition bonds to 9**

Number Connections © Rose Griffiths 2005
Harcourt Education Ltd

Make 9	Make 9	Make 9	Make 9
5	4	5	4

Number Connections © Rose Griffiths 2005
Harcourt Education Ltd

Make 9 game

sheet 2 of 2

Print on card if possible. Reusable.

Make 9	Make 9	Make 9	Make 9
5	4	6 six	3

Make 9	Make 9	Make 9	Make 9
6 six	3	6 six	3

Make 9	Make 9	Make 9	Make 9
7	2	7	2

Make 9	Make 9	Make 9	Make 9
8	1	9 nine	0

Joke shop

Name ______________________

Date ______________________

Draw the missing eyeballs.

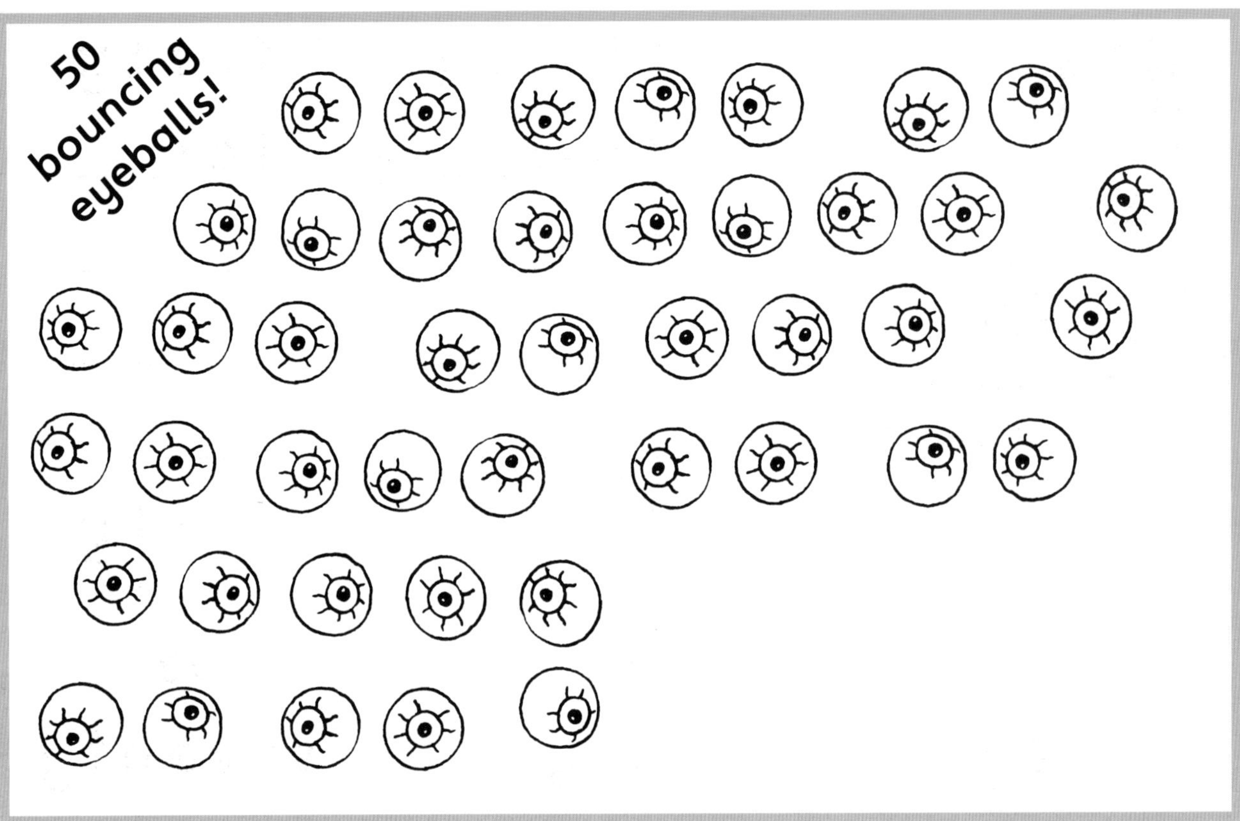

Draw the missing ants.

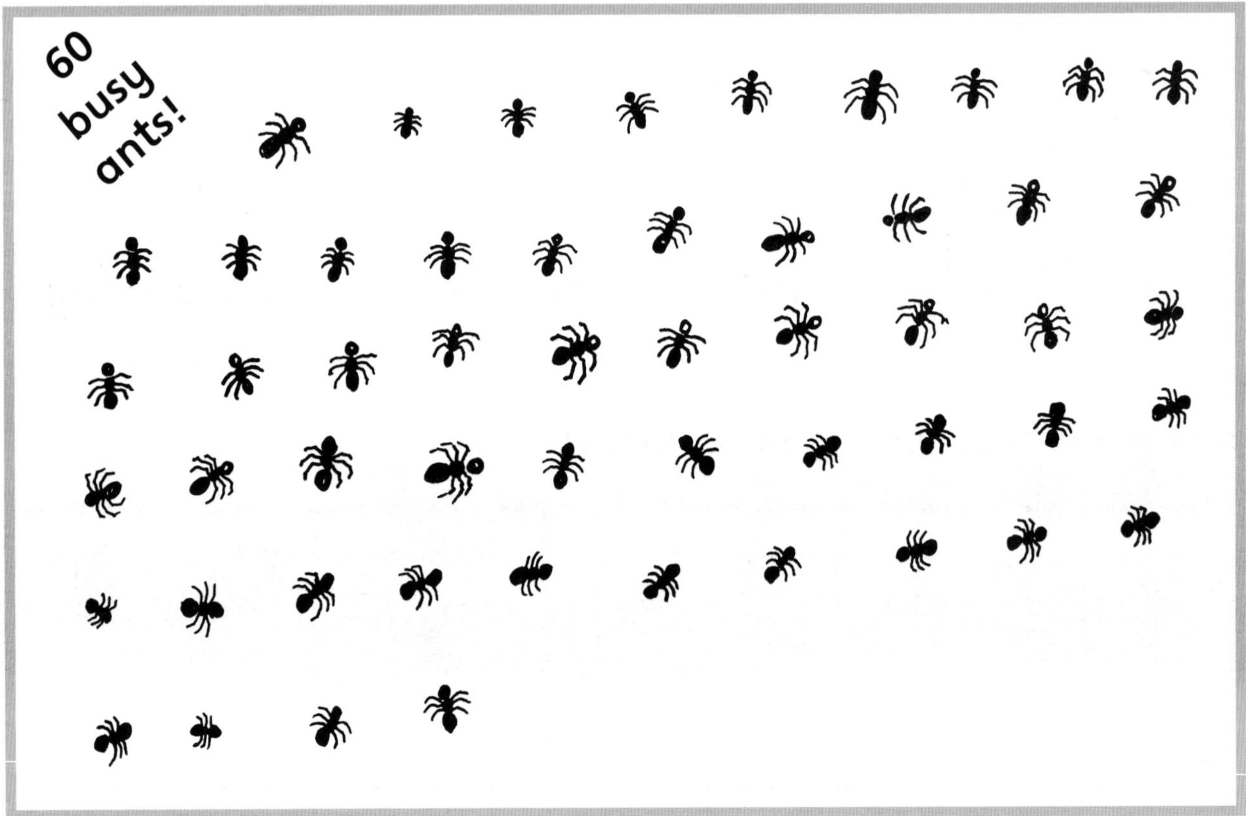

Tens and teens

Name ___________________

Date ___________________

30

t h i r t y
t h i r ___
t h _____
t _______
____
t h i r t y

40

f o r t y
f o r ___
f _______
____
f o r t y

50

f i f t y
f i f ___
f i _____
f _______
____
f i f t y

60

s i x t y
s i x ___
s i _____
s _______
____
s i x t y

70

s e v e n t y
s e v e n ___
s e v _____
s _______
____
s e v e n t y

Mixed up numbers! Sort them out.

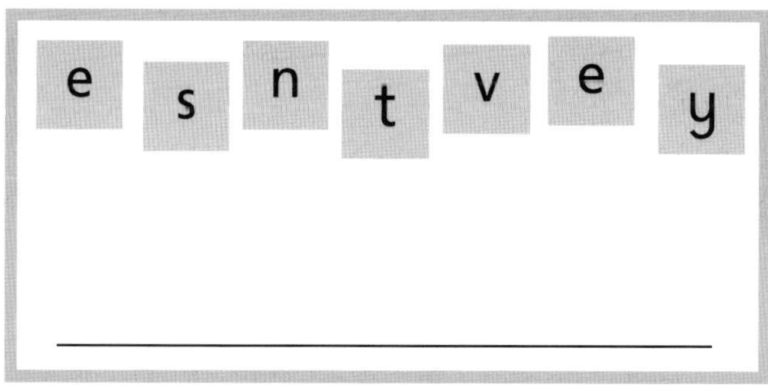

____

____

Tens and teens

Name ______________________

Date ______________________

Draw tens and ones.
Write the number.

thirteen

13

thirty

30

seventy

forty

fifteen

fourteen

seventeen

sixty

Write these numbers as words.

50 ______________________

16 ______________________

What comes next?

Name ______________________

Date ______________________

Join the dots.
They go up in twos.

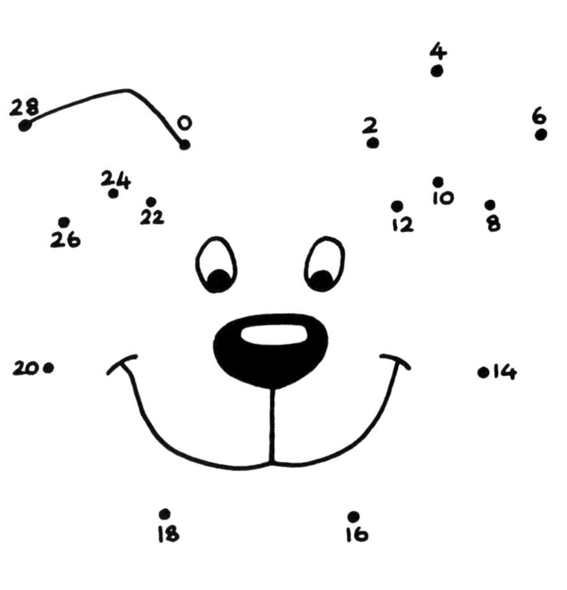

Join the dots.
They go up in fives.

Join the dots.
They go up in tens.

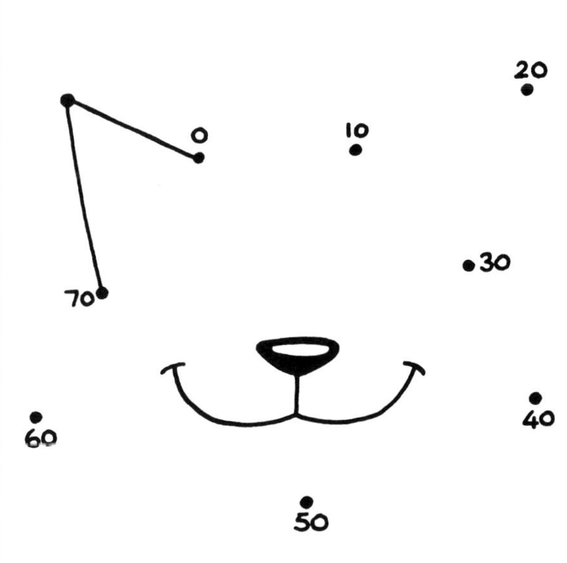

Draw eyes and whiskers.

Join the dots.
They go __down__ in fives.

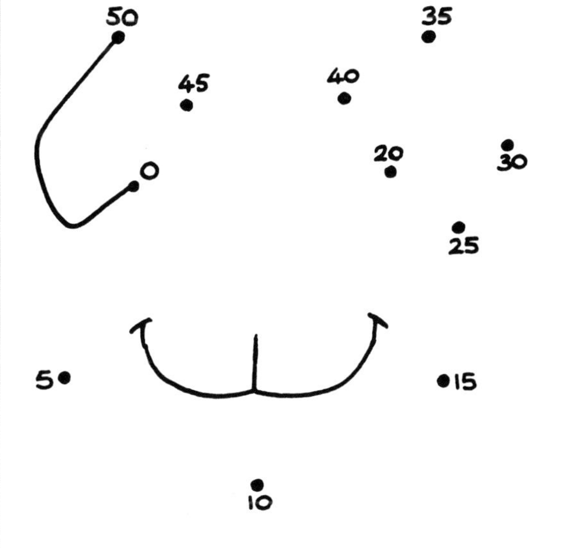

Draw eyes and a nose.

Multiples of 2, 5 and 10 ◀ Blue Pupil Book Part 3 pages 72 and 73 ▶ Copymaster B66

Number Connections © Rose Griffiths 2005
Harcourt Education Ltd

What comes next?

Name ______________________

Date ______________________

Fill in the missing numbers.

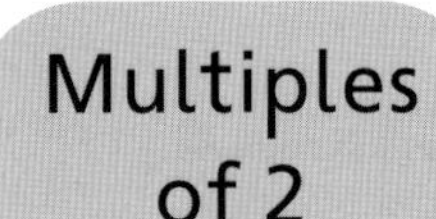

0
2
4
6
8
10

16

22

34

38

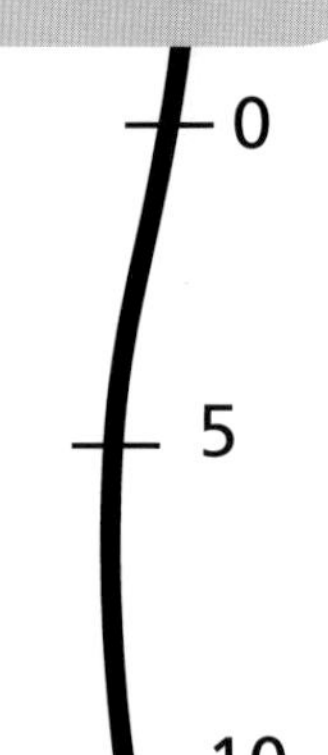

0

5

10

40

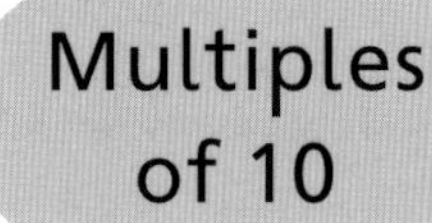

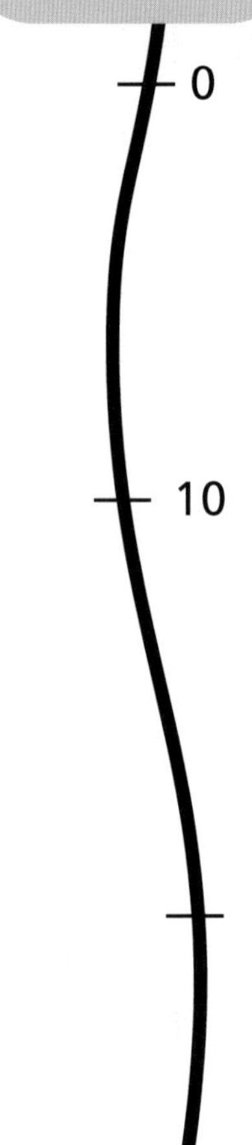

Multiples of 2, 5 and 10 ◄ Blue Pupil Book Part 3 pages 72 and 73

Number Connections © Rose Griffiths 2005
Harcourt Education Ltd

Footballs

Name _________________________

Date _________________________

3 + ☐ = 10

5 + ☐ = 10

4 + ☐ = 10

2 + ☐ = 10

6 + ☐ = 10

9 + ☐ = 10

7 + ☐ = 10

8 + ☐ = 10

1 + ☐ = 10

☐ + 1 = 10

☐ + 0 = 10

☐ + 4 = 10

☐ + 5 = 10

☐ + 7 = 10

☐ + 2 = 10

☐ + 8 = 10

☐ + 6 = 10

☐ + 3 = 10

6 + ☐ = 10

☐ + 6 = 10

3 + ☐ = 10

☐ + 3 = 10

8 + ☐ = 10

☐ + 8 = 10

1 + ☐ = 10

☐ + 1 = 10

Footballs

Name ________________________

Date ________________________

10 – 2 = ___	10 – 7 = ___	10 – 1 = ___
10 – 3 = ___	10 – 0 = ___	10 – 8 = ___
10 – 9 = ___	10 – 5 = ___	10 – 4 = ___
10 – 4 = ___	10 – 6 = ___	10 – 3 = ___
10 – 8 = ___	10 – 2 = ___	10 – 7 = ___

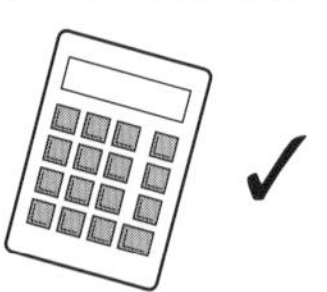 ✓ or ✗

How much change?

How much change?

How much change?

How much change?

Easier adding

Name ___________________

Date ___________________

10

$4 + 9 + 1 =$ __14__ $6 + 15 + 4 =$ ____

$3 + 5 + 7 =$ ____ $8 + 11 + 2 =$ ____

$2 + 8 + 9 =$ ____ $5 + 6 + 7 =$ ____

$13 + 4 + 6 =$ ____ $9 + 1 + 13 =$ ____

$5 + 9 + 5 =$ ____ $7 + 9 + 3 =$ ____

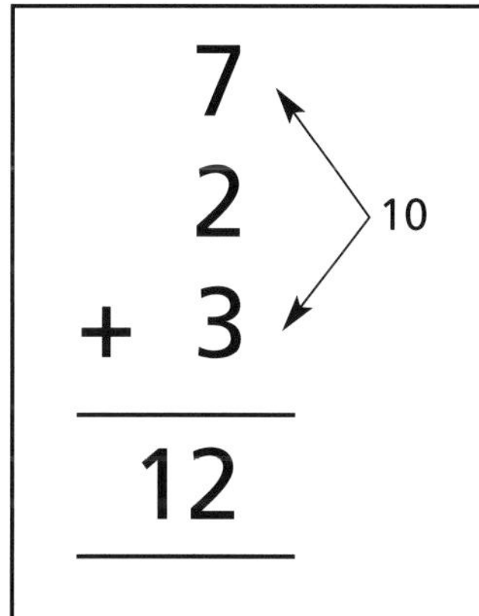

$\begin{array}{r} 4 \\ 8 \\ + 6 \\ \hline \end{array}$ $\begin{array}{r} 12 \\ 3 \\ + 7 \\ \hline \end{array}$ $\begin{array}{r} 6 \\ 11 \\ + 4 \\ \hline \end{array}$ $\begin{array}{r} 8 \\ 14 \\ + 2 \\ \hline \end{array}$

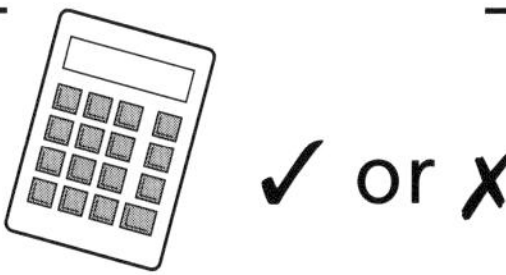 ✓ or ✗

Easier adding

Name _______________________

Date _______________________

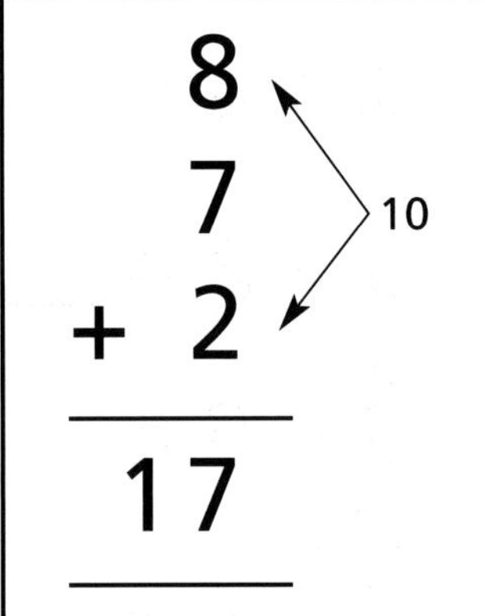

7 5 + 5	2 15 + 7	13 9 + 1	5 7 + 3
8 6 + 2	7 12 + 3	14 3 + 6	11 5 + 9
5 15 + 5	6 8 + 4	9 1 + 9	13 5 + 8

$4 + 2 + 8 + 6 =$ _____ $3 + 1 + 7 + 9 =$ _____

Speedy sums Ⓖ

1 2 3 minute test

Name ___________________

Date ___________________

5 + 4 = ____   9 − 5 = ____   4 + 4 = ____

2 + 8 = ____   7 − 6 = ____   5 − 2 = ____

3 + 3 = ____   4 − 4 = ____   6 + 3 = ____

8 + 0 = ____   10 − 8 = ____   9 − 7 = ____

1 + 6 = ____   7 − 3 = ____   5 + 5 = ____

6 + 4 = ____   8 − 3 = ____   8 − 5 = ____

3 + 7 = ____   2 − 1 = ____    Score: ____

Speedy sums Ⓗ

1 2 3 minute test

Name ___________________

Date ___________________

9 + 1 = ____   6 − 1 = ____   4 + 6 = ____

2 + 6 = ____   8 − 5 = ____   9 − 2 = ____

5 + 2 = ____   7 − 2 = ____   0 + 10 = ____

3 + 3 = ____   10 − 2 = ____   7 − 3 = ____

7 + 3 = ____   9 − 8 = ____   3 + 6 = ____

4 + 3 = ____   5 − 5 = ____   10 − 5 = ____

5 + 4 = ____   8 − 4 = ____    Score: ____

Adding up

Name _______________________

Date _______________________

15 + 19

15 + 19 =

23 + 16

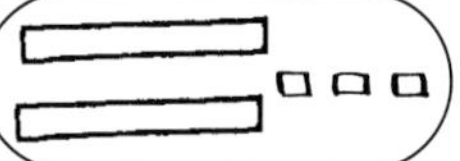 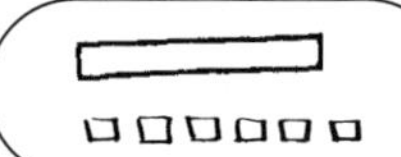

23 + 16 =

16 + 12

16 + 12 =

13 + 27

13 + 27 =

Draw tens and ones. 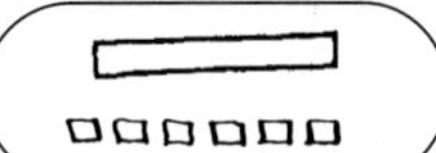

13 + 20

13 + 20 =

17 + 9

17 + 9 =

14 + 14

14 + 14 =

18 + 5

18 + 5 =

Adding up

Name ___________________

Date ___________________

25 + 13	19 + 11
25 + 13 =	19 + 11 =

16 + 16	17 + 14
16 + 16 =	17 + 14 =

Now try these. You can choose how to do them.

15 + 23	21 + 16
7 + 25	20 + 20

Three times table

Name ______________________

Date ______________________

Cut out the eleven tables facts.
Fold along the dotted line and glue flat.

		5 × 3	15
0 × 3	0	6 × 3	18
1 × 3	3	7 × 3	21
2 × 3	6 (six)	8 × 3	24
3 × 3	9 (nine)	9 × 3	27
4 × 3	12	10 × 3	30

Three times table

Name _______________________

Date _______________________

Fill in the missing numbers. Check with a calculator.

$10 \times 3 = \boxed{}$

$3 \times 10 = \boxed{}$

$4 \times 3 = \boxed{}$

$3 \times 4 = \boxed{}$

$7 \times 3 = \boxed{}$

$3 \times 7 = \boxed{}$

$5 \times 3 = \boxed{}$

$3 \times 5 = \boxed{}$

$10 \times 3 = \boxed{}$

$\boxed{} \times 3 = 27$

$8 \times 3 = \boxed{}$

$7 \times 3 = \boxed{}$

$\boxed{} \times 3 = 18$

$\boxed{} \times 3 = 15$

$4 \times 3 = \boxed{}$

$\boxed{} \times 3 = 9$

$2 \times 3 = \boxed{}$

$1 \times 3 = \boxed{}$

$0 \times 3 = \boxed{}$

____

____

Fifty pences

Name _______________________

Date _______________________

Draw the missing coins:  **or** .

£1.50

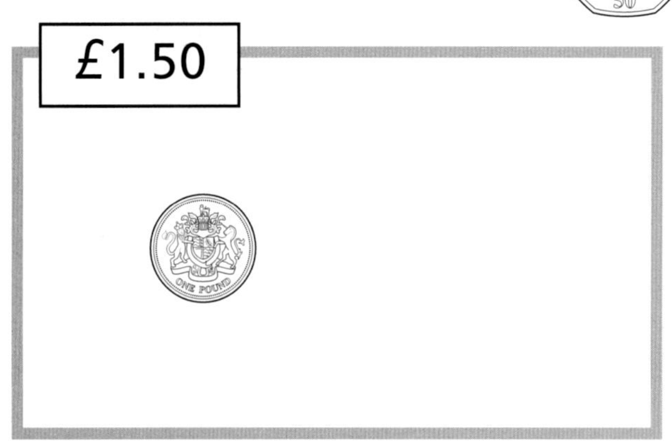

£3.50

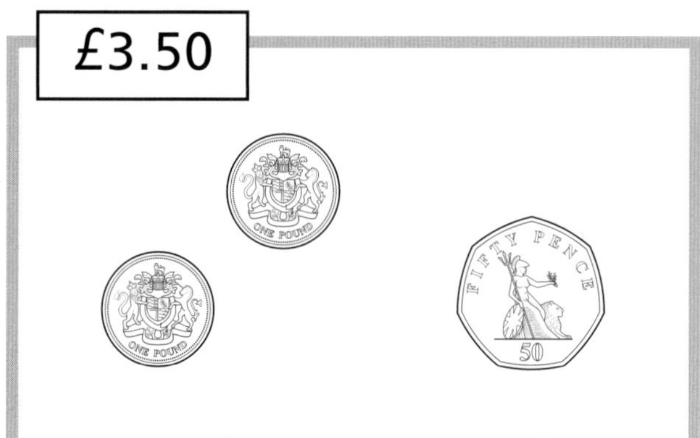

£5.00

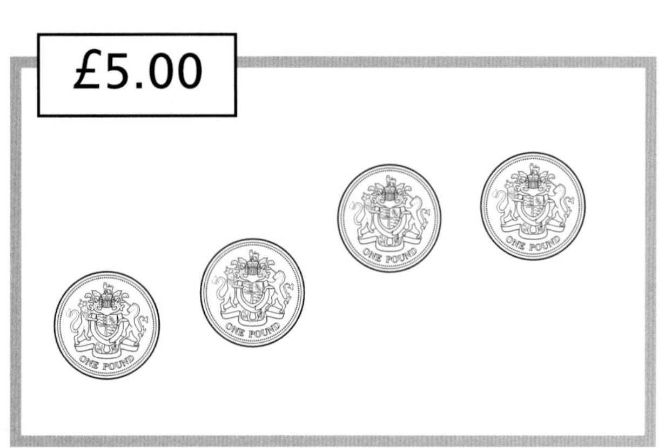

£1.50

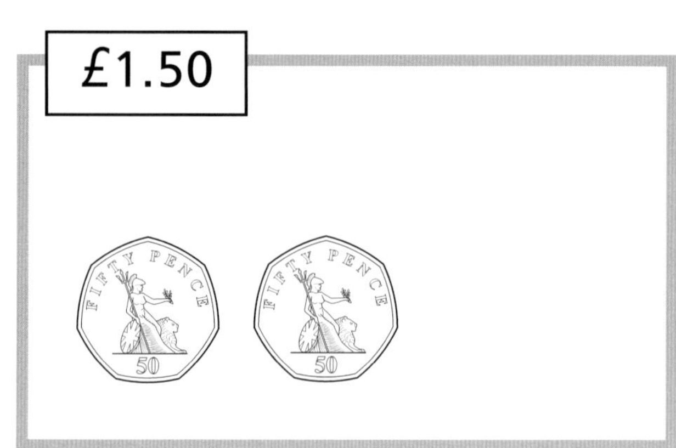

How much money is in each box?

£

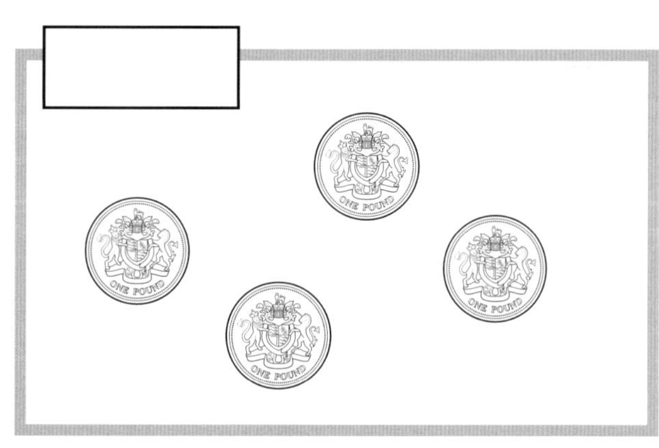

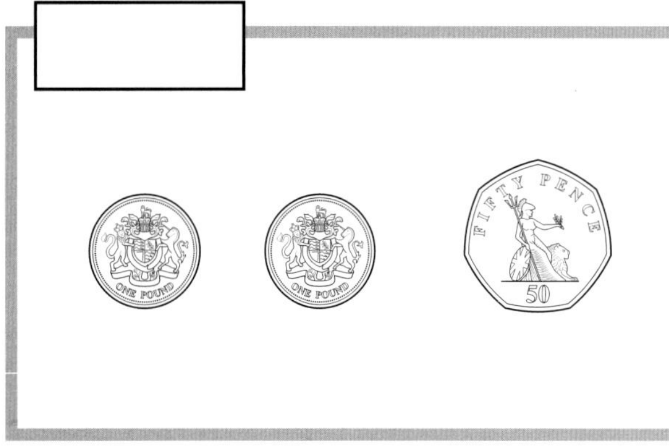

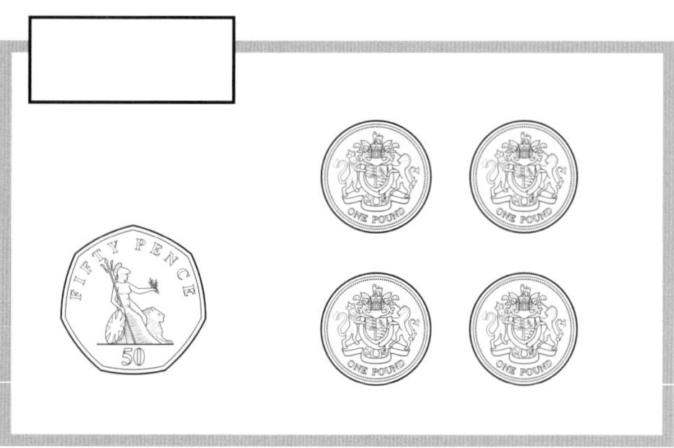

Fifty pences

Name _______________________

Date _______________________

Use coins. 

Draw 4 different ways of making 50p.

50p

50p

50p

50p

Dog's toys

Name ________________________

Date ________________________

Squeaky toys
£2 each

How many toys for £4?

Put the pounds in twos.

$4 \div 2 =$ ________

How many toys for £6?

$6 \div 2 =$ ________

How many toys for £8?

$8 \div 2 =$ ________

How many toys for £10?

$10 \div 2 =$ ________

How many toys for £12?

      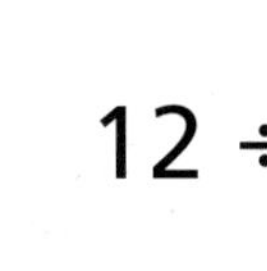

$12 \div 2 =$ ________

Multiplication and division by 2   ◀ Blue Pupil Book Part 3 pages 86 and 87
▶ Copymaster B79

Dog's toys

Name _______________________

Date _______________________

Squeaky toys
£2 each

Multiplying

How much for 10 toys?

1 0 × 2 = ___

9 × 2 = ___

8 × 2 = ___

7 × 2 = ___

6 × 2 = ___

5 × 2 = ___

4 × 2 = ___

3 × 2 = ___

2 × 2 = ___

1 × 2 = ___

Dividing

How many toys for £20?

2 0 ÷ 2 = ___

1 8 ÷ 2 = ___

1 6 ÷ 2 = ___

1 4 ÷ 2 = ___

1 2 ÷ 2 = ___

1 0 ÷ 2 = ___

8 ÷ 2 = ___

6 ÷ 2 = ___

4 ÷ 2 = ___

2 ÷ 2 = ___

Swimming

Name ___________________

Date ___________________

Fill in the missing numbers.

10	20		
9		29	
8		28	
	17		37
6	16		36
5		25	35
	14	24	
3	13		33
2		22	32
1	11	21	

Now do the questions on Pupil Book page 89.

Swimming

Name ___________________

Date ___________________

Fill in the missing letters.

1st	2nd	3rd	4th
first	second	third	fourth
fir___	sec_____	th___	four___
f_____	s_______	t_____	f_______
f_____	s_______	t_____	f_______
_____	_______	_____	_______
first	second	third	fourth

We had a swimming race.

Finish writing these.

Dog food

Name _______________________

Date _______________________

$4 \times 2 =$ _____ $2 \times 4 =$ _____ $7 \times 2 =$ _____

$2 \times 2 =$ _____ $6 \times 2 =$ _____ $9 \times 2 =$ _____

$8 \times 2 =$ _____ $2 \times 7 =$ _____ $2 \times 3 =$ _____

$2 \times 5 =$ _____ $10 \times 2 =$ _____ $5 \times 2 =$ _____

$3 \times 2 =$ _____ $2 \times 9 =$ _____ $2 \times 10 =$ _____

$2 \times 6 =$ _____ $2 \times 8 =$ _____ $1 \times 2 =$ _____

$\square \times 2 = 6$ $6 \div 2 =$ ___	$\square \times 2 = 14$ $14 \div 2 =$ ___
$\square \times 2 = 8$ $8 \div 2 =$ ___	$\square \times 2 = 18$ $18 \div 2 =$ ___

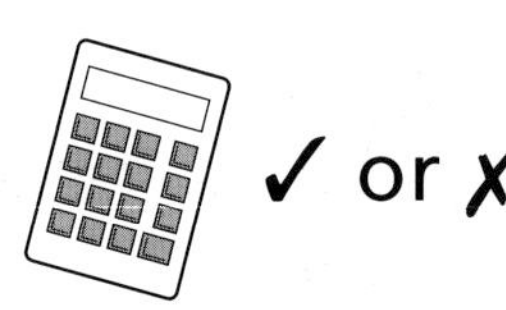 ✓ or ✗

Dog food

Name _______________________

Date _______________________

$\square \times 2 = 4$

$4 \div 2 = \underline{}$

$\square \times 2 = 10$

$10 \div 2 = \underline{}$

$\square \times 2 = 12$

$12 \div 2 = \underline{}$

$\square \times 2 = 16$

$16 \div 2 = \underline{}$

$8 \div 2 = \underline{}$	$6 \div 2 = \underline{}$	$8 \div 2 = \underline{}$
$4 \div 2 = \underline{}$	$16 \div 2 = \underline{}$	$14 \div 2 = \underline{}$
$10 \div 2 = \underline{}$	$4 \div 2 = \underline{}$	$12 \div 2 = \underline{}$
$14 \div 2 = \underline{}$	$20 \div 2 = \underline{}$	$16 \div 2 = \underline{}$
$2 \div 2 = \underline{}$	$18 \div 2 = \underline{}$	$10 \div 2 = \underline{}$
$12 \div 2 = \underline{}$	$2 \div 2 = \underline{}$	$18 \div 2 = \underline{}$

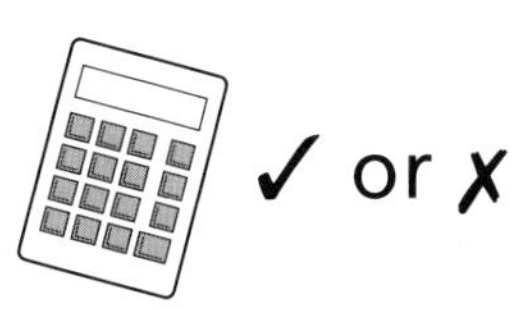 ✓ or ✗

Taking away

Name _______________________

Date _______________________

27 – 15

27 – 15 =

19 – 11

19 – 11 =

Cross out tens and ones.

36 – 13

36 – 13 =

28 – 22

28 – 22 =

18 – 7

18 – 7 =

34 – 14

34 – 14 =

27 – 16

27 – 16 =

29 – 14

29 – 14 =

Taking away

Name _______________________

Date _______________________

Use tens and ones. 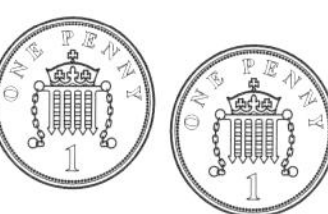

How much did each person have left?

He had ☐ left.

She had ☐ left.

She had ☐ left.

He had ☐ left.

She had ☐ left.

He had ☐ left.

He had ☐ left.

She had ☐ left.

Secret numbers

Name _______________________

Date _______________________

Use cards numbered 1 to 20. 

Which numbers belong in each box?

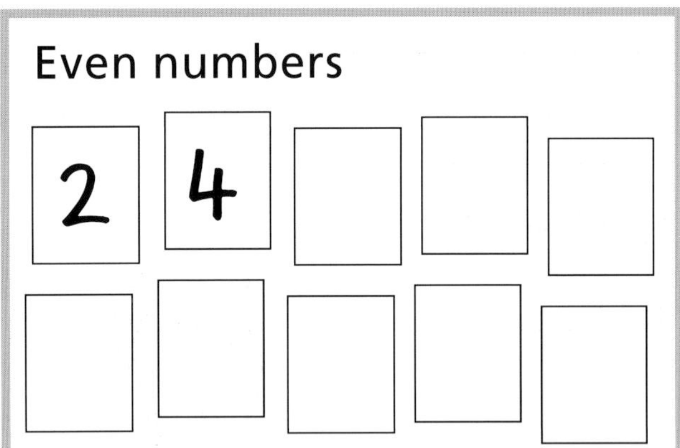

Even numbers

2　4

Odd numbers

1　3

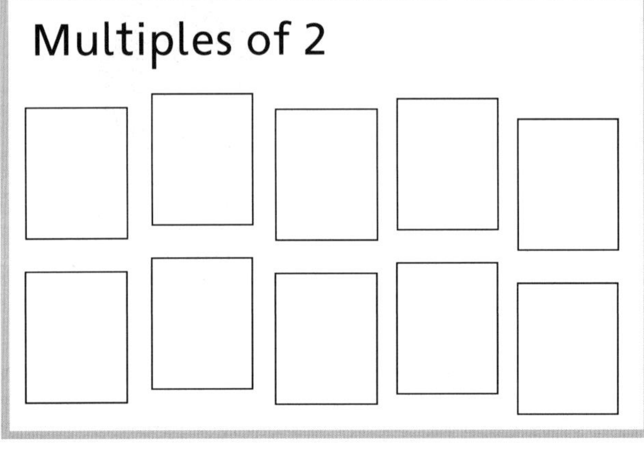

Multiples of 2

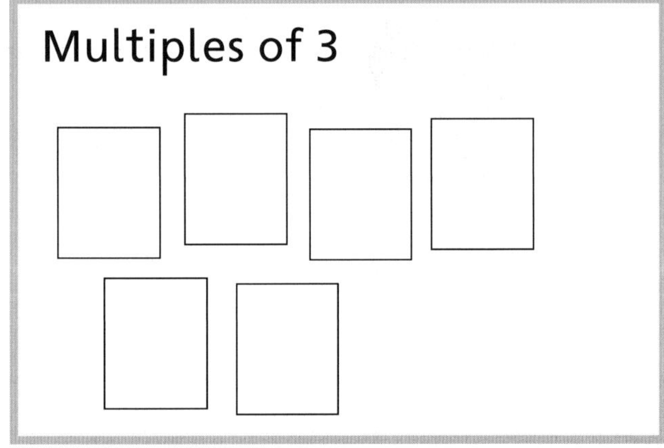

Multiples of 3

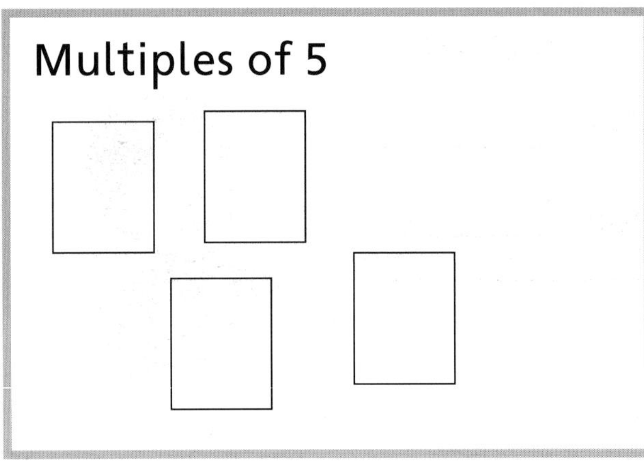

Multiples of 5

Multiples of 10

Secret numbers

Name ___________________

Date ___________________

Use cards numbered 1 to 20.
Make up puzzles. Give them to a friend.

These puzzles are for ___________________________

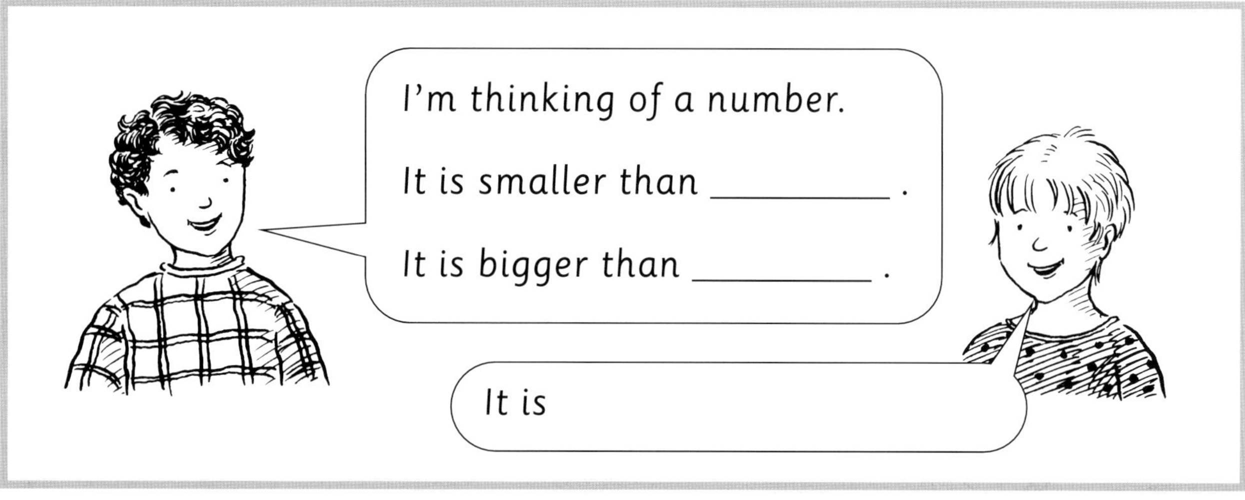

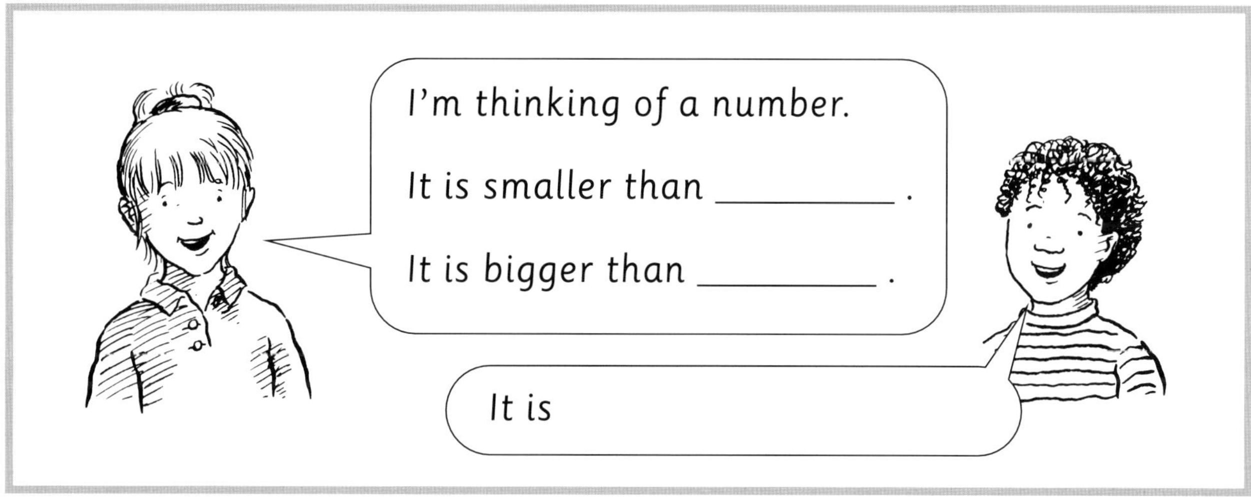

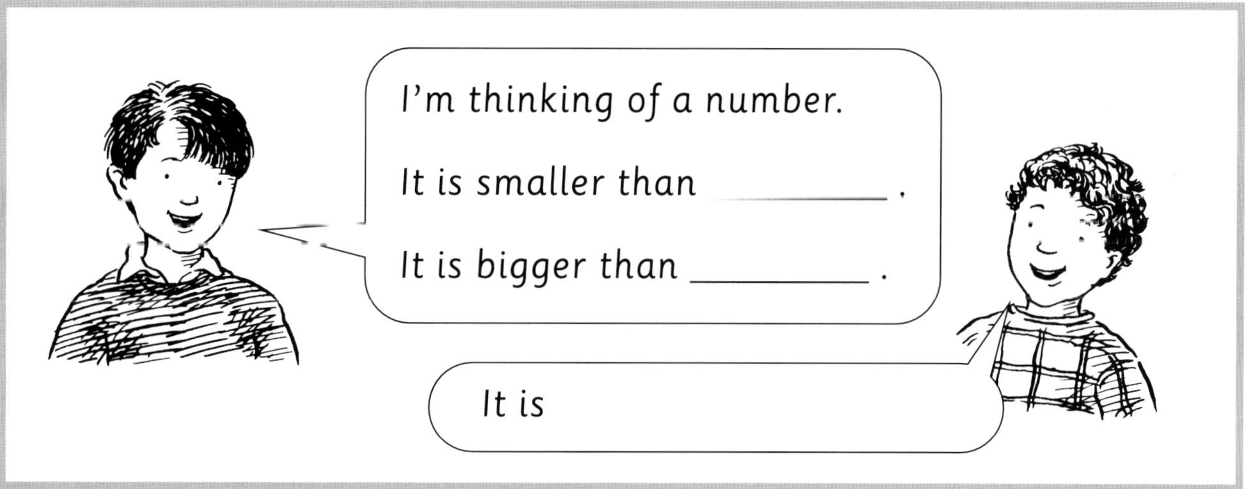

Tens and teens bingo

sheet 1 of 2

Print on card if possible. Reusable.
Cut out the instructions card, 4 bingo cards, and 10 number cards.
Store in a clear zip-top wallet or in an envelope. If possible, include 24 counters.

≈ **Tens and teens bingo** ≈

A game for 2, 3 or 4 people.

- **Before you start**
  You need a bingo card and 6 counters each.
  Shuffle the 10 number cards.
  Put them in a pile, face down.

- **How to play**

If that number is on their bingo card,
they cover it up with a counter.

- **Keep going until someone says 'Bingo' because
  they have covered up all their numbers.**

◄ Blue Pupil Book Part 3; **Saying and listening to tens and teen numbers**

Number Connections © Rose Griffiths 2005
Harcourt Education Ltd

seventy	Tens and teens bingo	sixty	Tens and teens bingo	fifty	Tens and teens bingo	forty	Tens and teens bingo	thirty	Tens and teens bingo
seventeen	Tens and teens bingo	sixteen	Tens and teens bingo	fifteen	Tens and teens bingo	fourteen	Tens and teens bingo	thirteen	Tens and teens bingo

Tens and teens bingo

sheet 2 of 2

Print on card if possible. Reusable.

GP

| Tens and teens bingo | 15 17 | 15 | 40 | |
| | 60 | | 70 | 13 |

Tens and teens bingo		15		
	17	40		
	14			
	30	60		

Tens and teens bingo	50			
	13	16		
	40			
	70	15		

Tens and teens bingo	70			
	16		50	
	30	17		
	14			

Number Connections © Rose Griffiths 2005
Harcourt Education Ltd

Sums which make 10

sheet 1 of 3

Print on card if possible. Reusable.
Cut out the instructions card, 3 sum cards and 40 number cards.
Store in a clear zip-top wallet or in an envelope.

≈ Sums which make 10 ≈

- **Before you start**
  Take a sum card each.
  Shuffle the number cards.
  Put them in a pile on the table, face down.

- **How to play**
  Can you make 4 different sums which make 10?

- **Keep going until you have all filled your cards.**
  Who finished first?

◄ Blue Pupil Book Part 3; **Addition and subtraction bonds to 10**

Number Connections © Rose Griffiths 2005
Harcourt Education Ltd

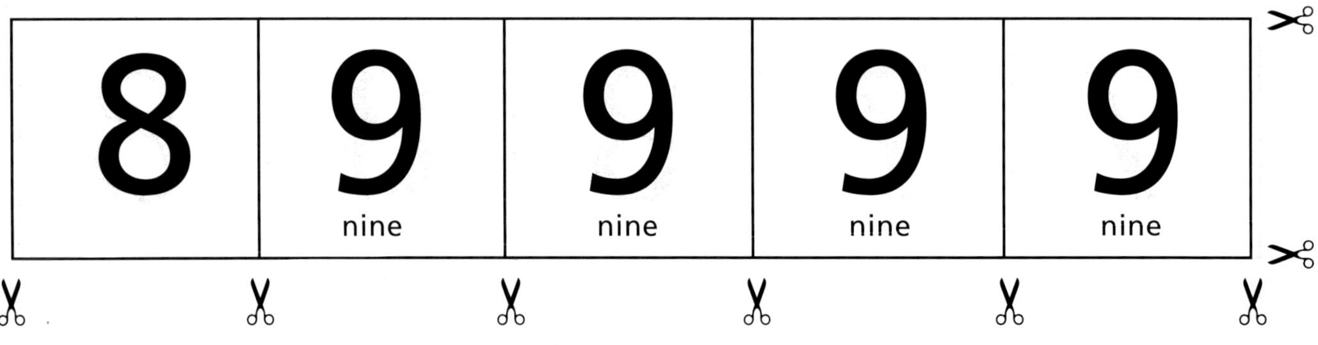

Sums which make 10

sheet 2 of 3

Print on card if possible. Reusable.

1	1	2	2	3
1	1	2	2	3
3	4	4	5	5
3	4	4	5	5
5	5	6 six	6 six	7
5	5	6 six	6 six	7
7	7	8	8	8

Number Connections © Rose Griffiths 2005
Harcourt Education Ltd

Sums which make 10

sheet 3 of 3

B92

T

Print 3 copies, on card if possible. Reusable.

GP

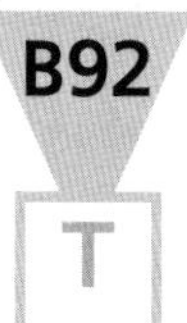

≈ Sums which make 10 ≈

☐ + ☐ = 10

☐ + ☐ = 10

☐ + ☐ = 10

☐ + ☐ = 10